NUMBER PALS

LEVEL B

Options
Publishing

Table of Contents

Number Sense and Operations

2

Algebra

Geometry

Measurement

Data Analysis

Count Tens and Ones

▶ **Count the blocks. Then fill in the blanks.**

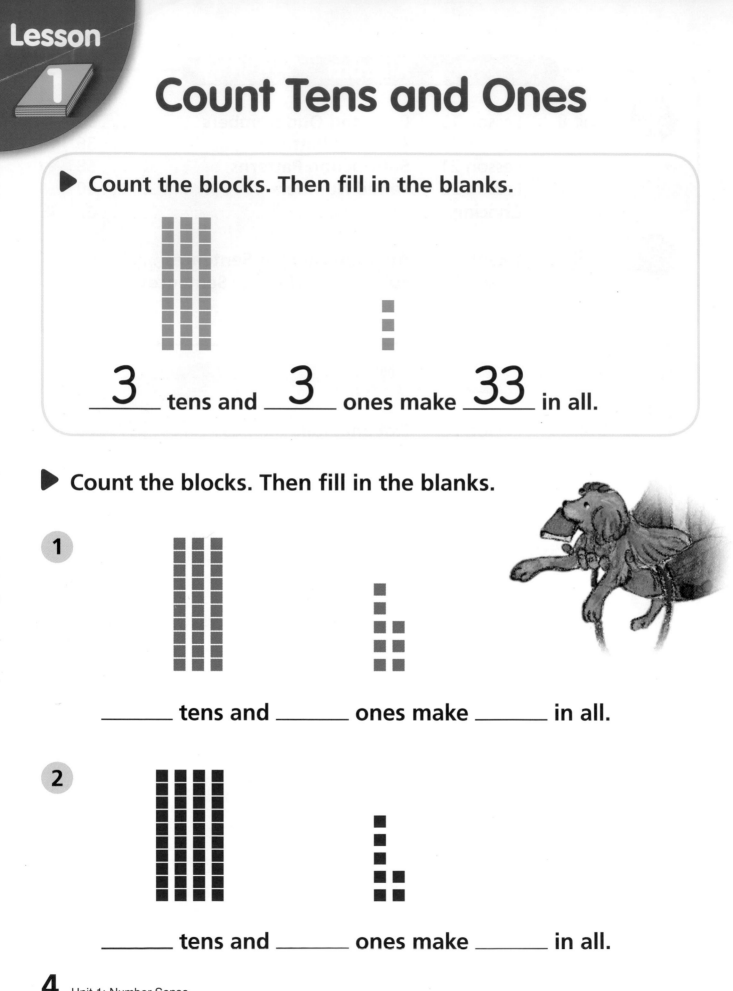

___3___ tens and ___3___ ones make ___33___ in all.

▶ **Count the blocks. Then fill in the blanks.**

1

_____ tens and _____ ones make _____ in all.

2

_____ tens and _____ ones make _____ in all.

3

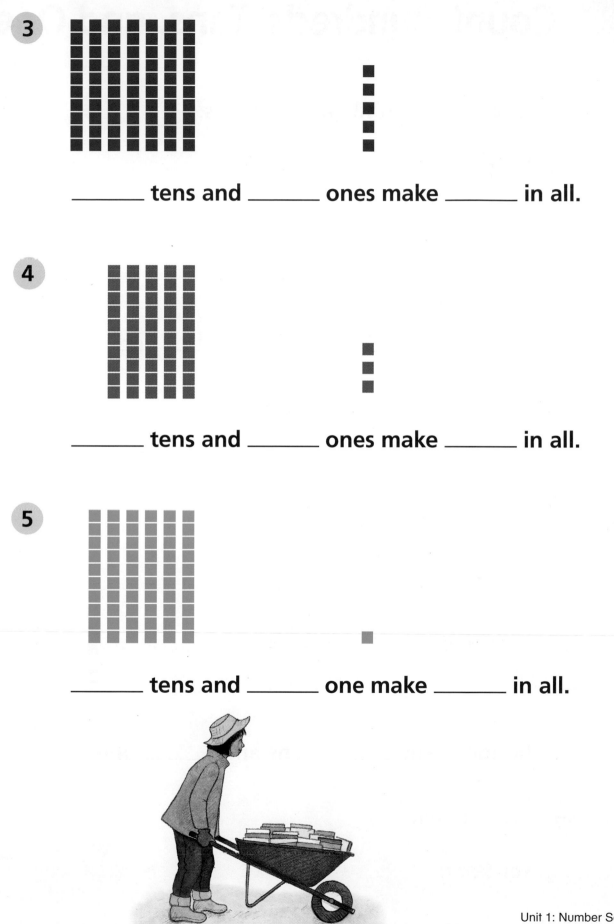

_____ tens and _____ ones make _____ in all.

4

_____ tens and _____ ones make _____ in all.

5

_____ tens and _____ one make _____ in all.

Count Hundreds, Tens, and Ones

▶ **Count the blocks. Then fill in the blanks.**

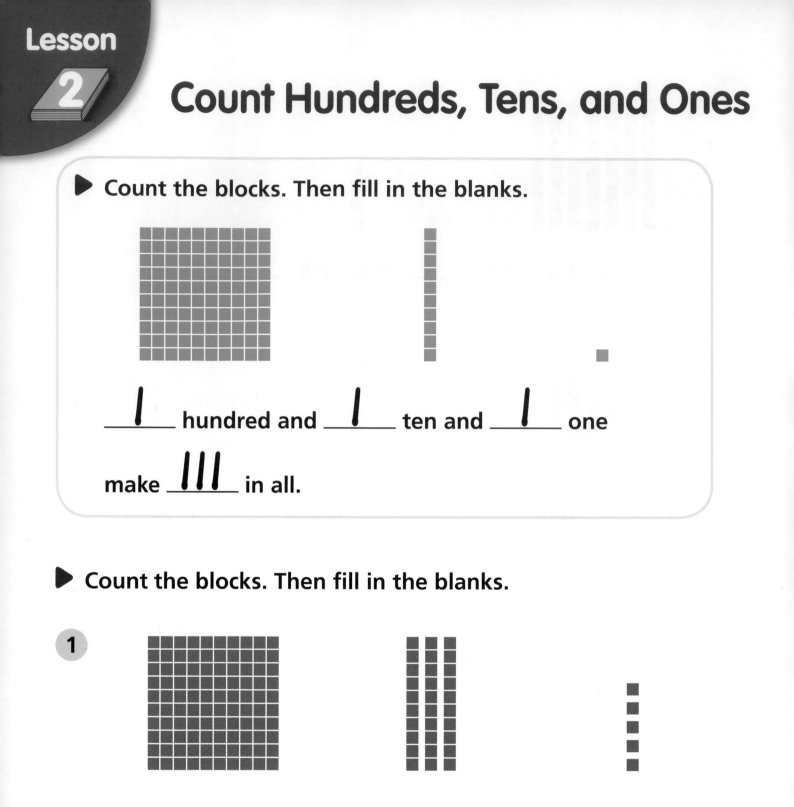

___/___ hundred and ___/___ ten and ___/___ one

make ___/ / / /___ in all.

▶ **Count the blocks. Then fill in the blanks.**

1

_____ hundred and _____ tens and _____ ones

make _____ in all.

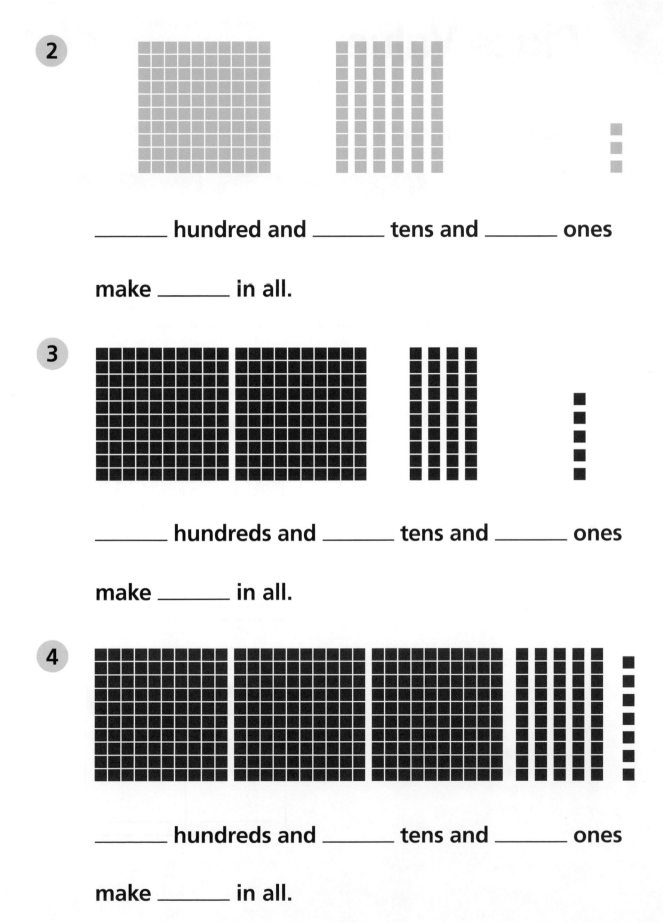

2

_____ hundred and _____ tens and _____ ones

make _____ in all.

3

_____ hundreds and _____ tens and _____ ones

make _____ in all.

4

_____ hundreds and _____ tens and _____ ones

make _____ in all.

Place Value

▶ **Count the blocks. Then fill in the chart.**

Hundreds	Tens	Ones
1	1	4

▶ **Count the blocks. Then fill in the chart.**

1

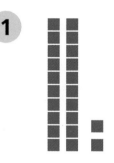

Tens	Ones

2

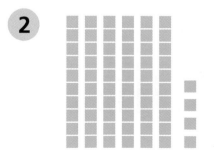

Tens	Ones

3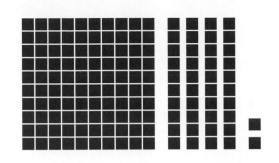

Hundreds	Tens	Ones
_____	_____	_____

4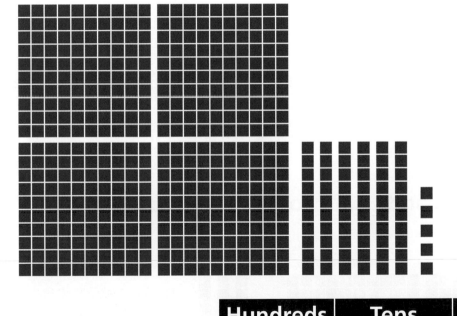

Hundreds	Tens	Ones
_____	_____	_____

Checkup

▶ **Count the blocks. Then fill in the blanks.**

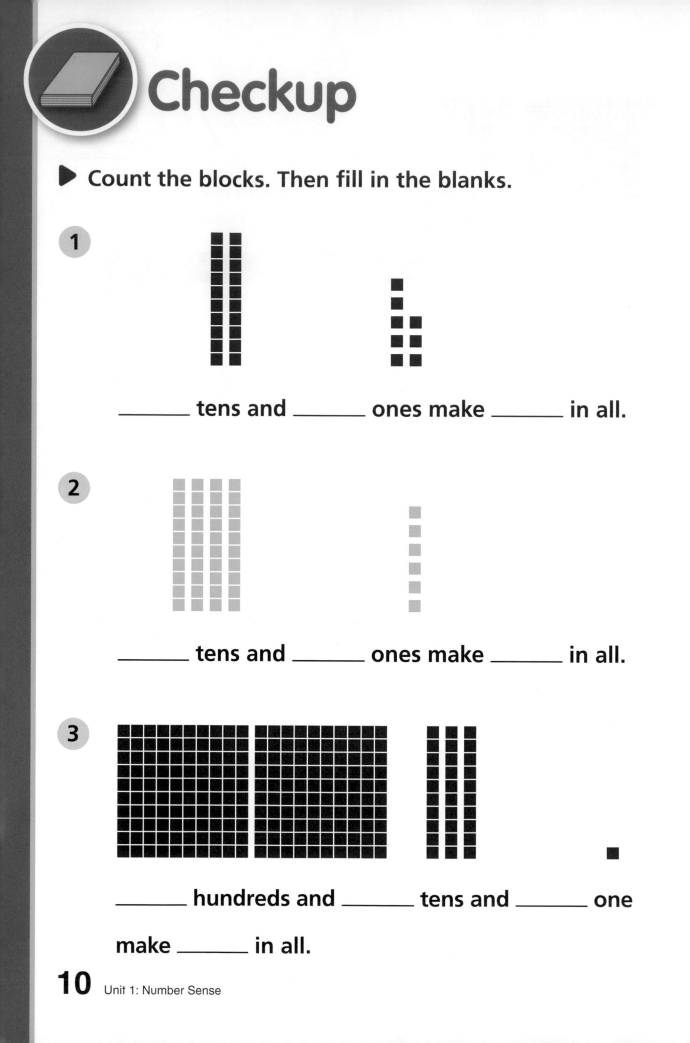

1

_____ tens and _____ ones make _____ in all.

2

_____ tens and _____ ones make _____ in all.

3

_____ hundreds and _____ tens and _____ one

make _____ in all.

▶ **Count the blocks. Then fill in the chart.**

4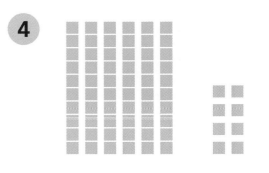

Tens	Ones
_____	_____

5

Tens	Ones
_____	_____

6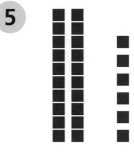

Hundreds	Tens	Ones
_____	_____	_____

Compare Numbers

You can compare numbers.

139 $=$ 139   139 is equal to 139

43 $<$ 53   43 is less than 53

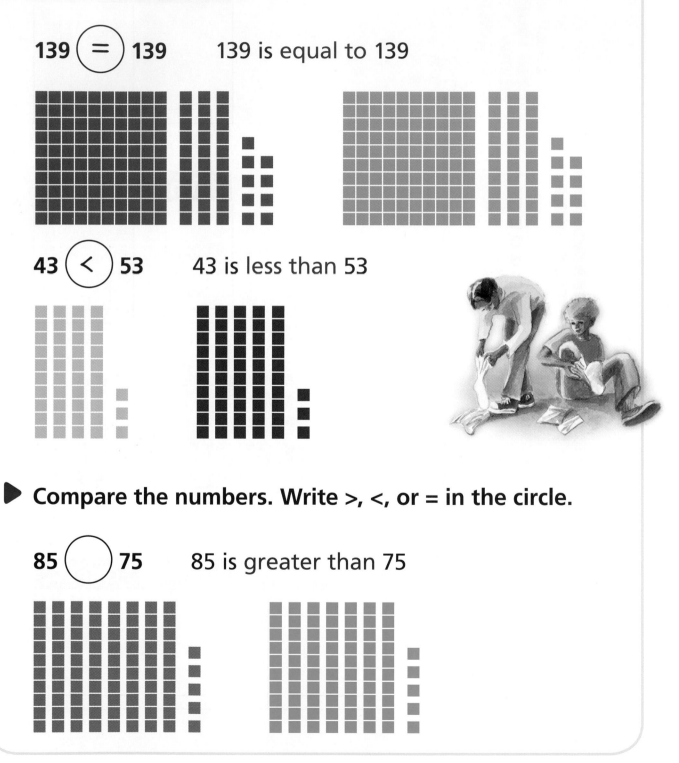

▶ **Compare the numbers. Write >, <, or = in the circle.**

85 ◯ 75   85 is greater than 75

Compare the numbers. Write >, <, or = in the circle.

1   185 ◯ 158

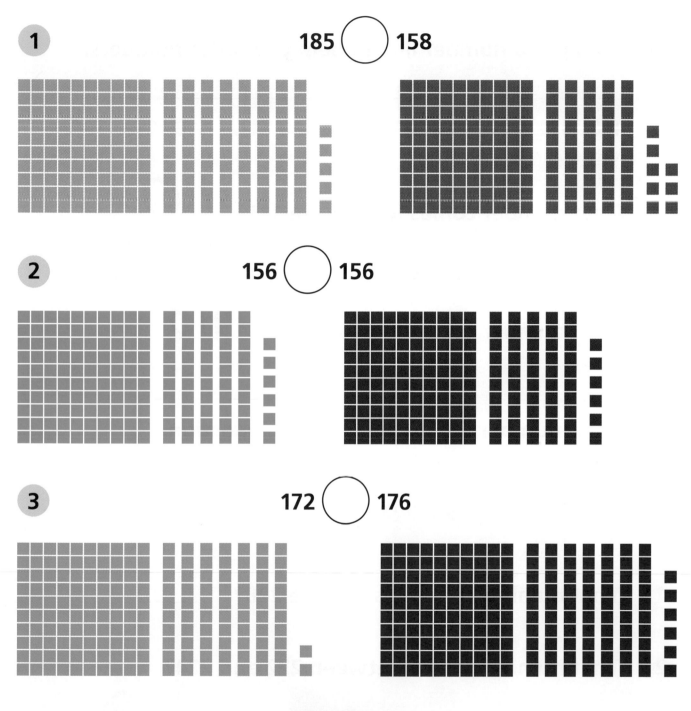

2   156 ◯ 156

3   172 ◯ 176

Order Numbers

You can use a number line to help you order numbers.

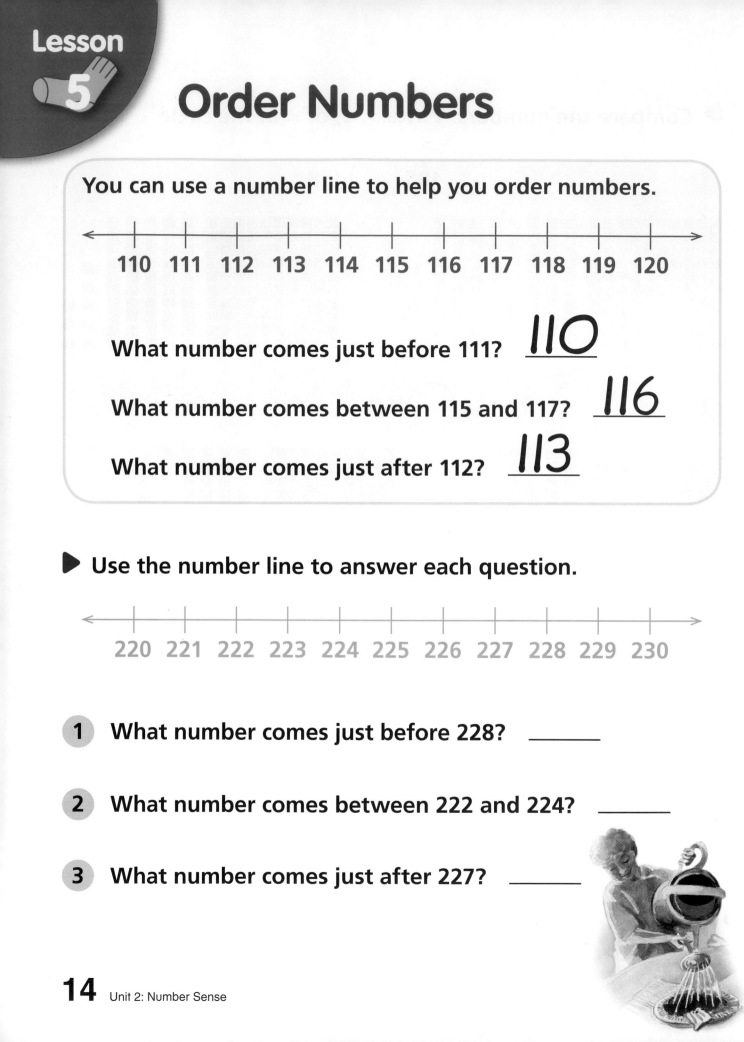

110 111 112 113 114 115 116 117 118 119 120

What number comes just before 111? __110__

What number comes between 115 and 117? __116__

What number comes just after 112? __113__

▶ Use the number line to answer each question.

220 221 222 223 224 225 226 227 228 229 230

1 What number comes just before 228? _____

2 What number comes between 222 and 224? _____

3 What number comes just after 227? _____

▶ **Use the number line to answer each question.**

$$250 \quad 251 \quad 252 \quad 253 \quad 254 \quad 255 \quad 256 \quad 257 \quad 258 \quad 259 \quad 260$$

4 What number comes just before 251? _____

5 What number comes between 253 and 255? _____

6 What number comes just after 256? _____

▶ **Use the number line to answer each question.**

$$300 \quad 301 \quad 302 \quad 303 \quad 304 \quad 305 \quad 306 \quad 307 \quad 308 \quad 309 \quad 310$$

7 What number comes just before 302? _____

8 What number comes between 304 and 306? _____

9 What number comes just after 309? _____

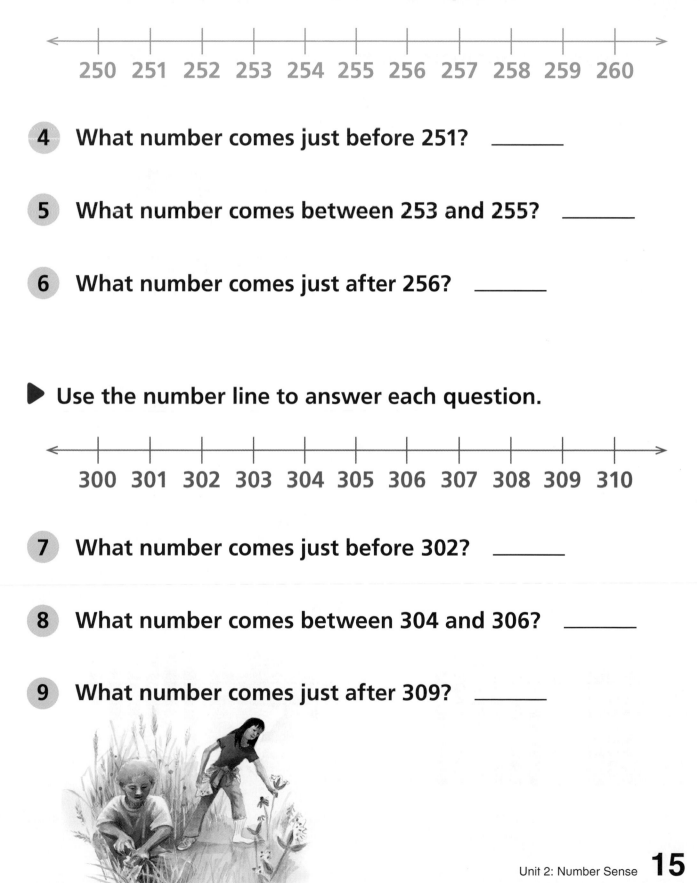

Checkup

▶ **Compare the numbers. Write >, <, or = in the circle.**

1   173 ◯ 137

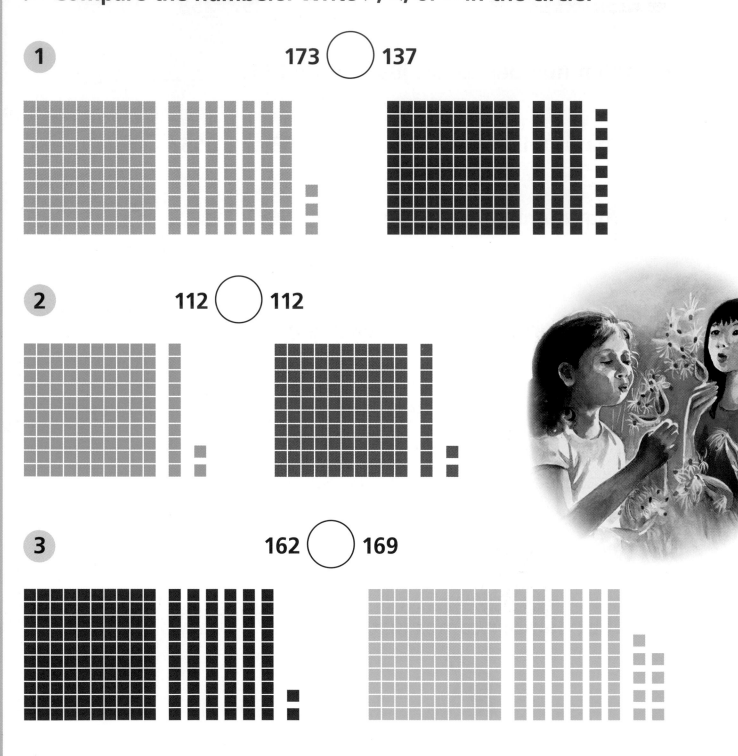

2   112 ◯ 112

3   162 ◯ 169

▶ **Use the number line to answer each question.**

<---+----+----+----+----+----+----+----+----+----+----+--->
   90   91   92   93   94   95   96   97   98   99 100

4 What number comes just before 98? _____

5 What number comes between 92 and 94? _____

6 What number comes just after 99? _____

▶ **Use the number line to answer each question.**

<---+----+----+----+----+----+----+----+----+----+----+--->
 200 201 202 203 204 205 206 207 208 209 210

7 What number comes just before 204? _____

8 What number comes between 205 and 207? _____

9 What number comes just after 208? _____

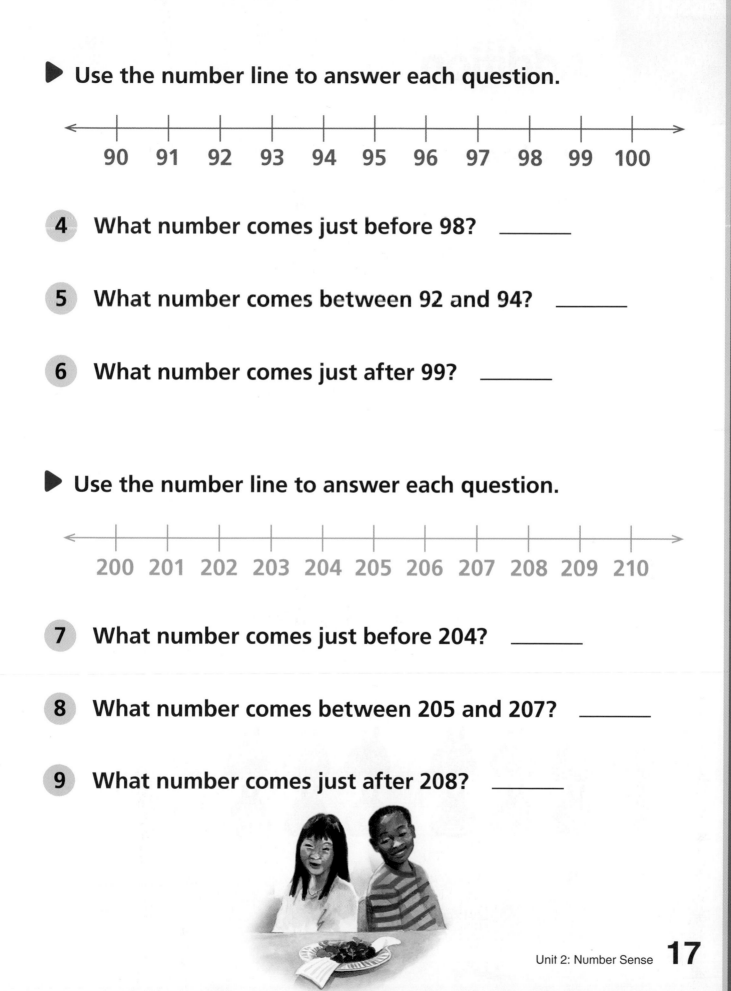

Addition

▶ **Add to find how many animals there are in all.**

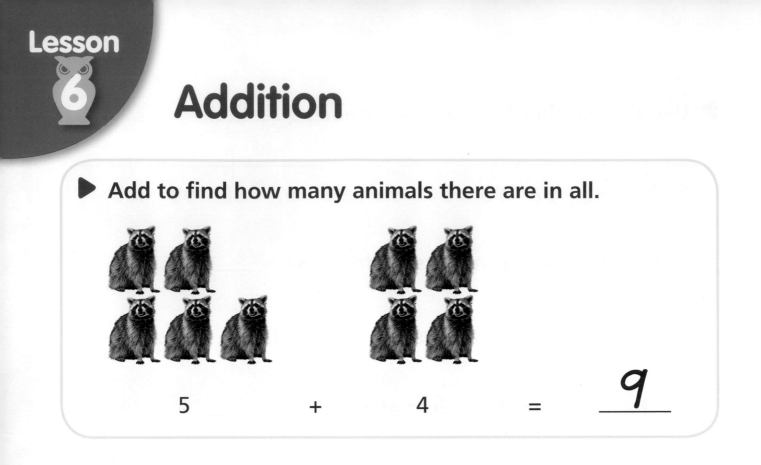

5   +   4   =   _9_

▶ **Add to find how many animals there are in all.**

1

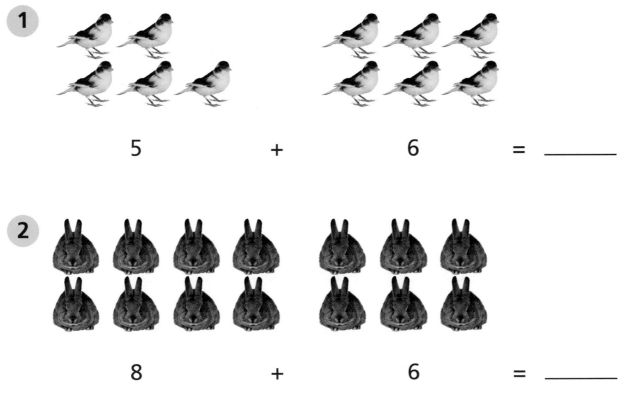

5   +   6   = _____

2

8   +   6   = _____

3

12   +   3   = _____

4

10   +   4   = _____

5

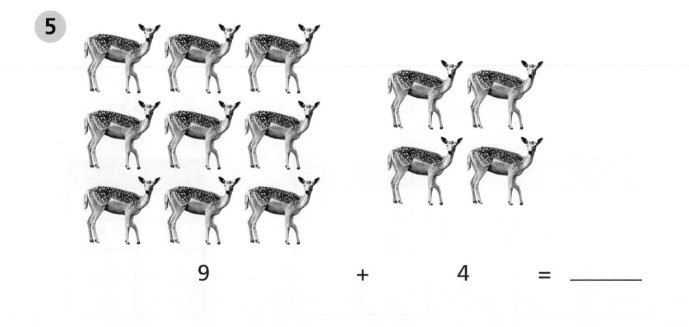

9   +   4   = _____

Add Tens and Ones

▶ **Add the numbers. 15 + 14 = ?**

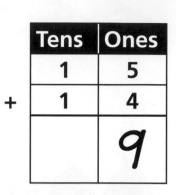

First, add the ones.
5 + 4 = 9.
Write 9 in the
ones place.

	Tens	Ones
	1	5
+	1	4
		9

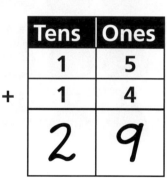

Then, add the tens.
1 + 1 = 2.
Write 2 in the
tens place.

15 + 14 = __29__

	Tens	Ones
	1	5
+	1	4
	2	9

▶ **Add the numbers.**

1   13 + 12 = ?

	Tens	Ones
	1	3
+	1	2

2   43 + 25 = ?

	Tens	Ones
	4	3
+	2	5

3   22 + 35 = ?

	Tens	Ones
	2	2
+	3	5

4 25 + 12 = ?

Tens	Ones
2	5
+ 1	2

5 31 + 41 = ?

Tens	Ones
3	1
+ 4	1

6 54 + 45 = ?

Tens	Ones
5	4
+ 4	5

7 27 + 62 = ?

Tens	Ones
2	7
+ 6	2

8 51 + 12 = ?

Tens	Ones
5	1
+ 1	2

9 55 + 32 = ?

Tens	Ones
5	5
+ 3	2

▶ **Add.**

10
```
   23
 + 12
_____
```

11
```
   45
 +34
_____
```

12
```
   63
  +32
_____
```

13
```
   72
  +26
_____
```

Addition with Regrouping

▶ **Regroup to add. 29 + 13 = ?**

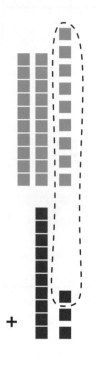

First, add the ones.
9 + 3 = 12.
Regroup 12 as
1 ten and 2 ones.
Write 2 in the
ones place.

Tens	Ones
1	
2	9
1	3
	2

+

Then, add the tens.
1 + 2 + 1 = 4.
Write 4 in the tens
place.

Tens	Ones
1	
2	9
1	3
4	2

+

29 + 13 = _____

▶ **Regroup to add.**

1 26 + 35 = ?

Tens	Ones
2	6
+ 3	5

2 36 + 35 = ?

Tens	Ones
3	6
+ 3	5

3 37 + 15 = ?

Tens	Ones
3	7
+ 1	5

4 59 + 16 = ?

Tens	Ones
5	9
+ 1	6

5 39 + 23 = ?

Tens	Ones
3	9
+ 2	3

6 79 + 16 = ?

Tens	Ones
7	9
+ 1	6

▶ **Add.**

7
```
  46
+ 38
────
```

8
```
  54
+ 26
────
```

9
```
  64
+ 27
────
```

10
```
  82
+ 18
────
```

Checkup

▶ **Add to find how many animals there are in all.**

1

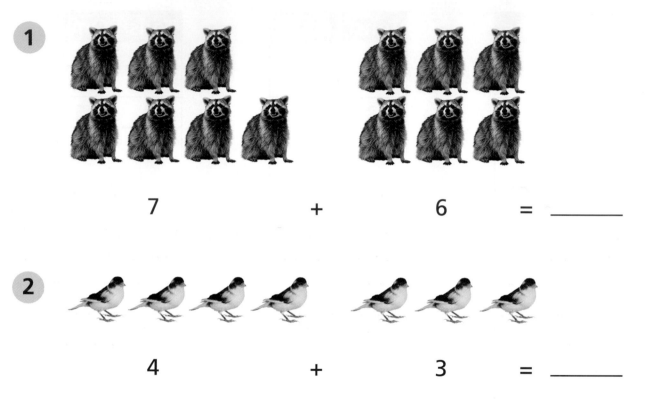

7   +   6   = _____

2

4   +   3   = _____

▶ **Add the numbers.**

3 26 + 12 = ?

Tens	Ones
2	6
+ 1	2

4 33 + 36 = ?

Tens	Ones
3	3
+ 3	6

5 41 + 15 = ?

Tens	Ones
4	1
+ 1	5

▶ Add.

6   24
       +12

7   33
       +44

8   42
       +53

9   51
       +24

▶ Regroup to add.

10 18 + 14 = ?

Tens	Ones
1	8
+ 1	4

11 27 + 38 = ?

Tens	Ones
2	7
+ 3	8

12 56 + 35 = ?

Tens	Ones
5	6
+ 3	5

▶ Add.

13   24
       +18

14   28
       +44

15   57
       +24

16   53
       +18

Subtraction

▶ **Subtract.**

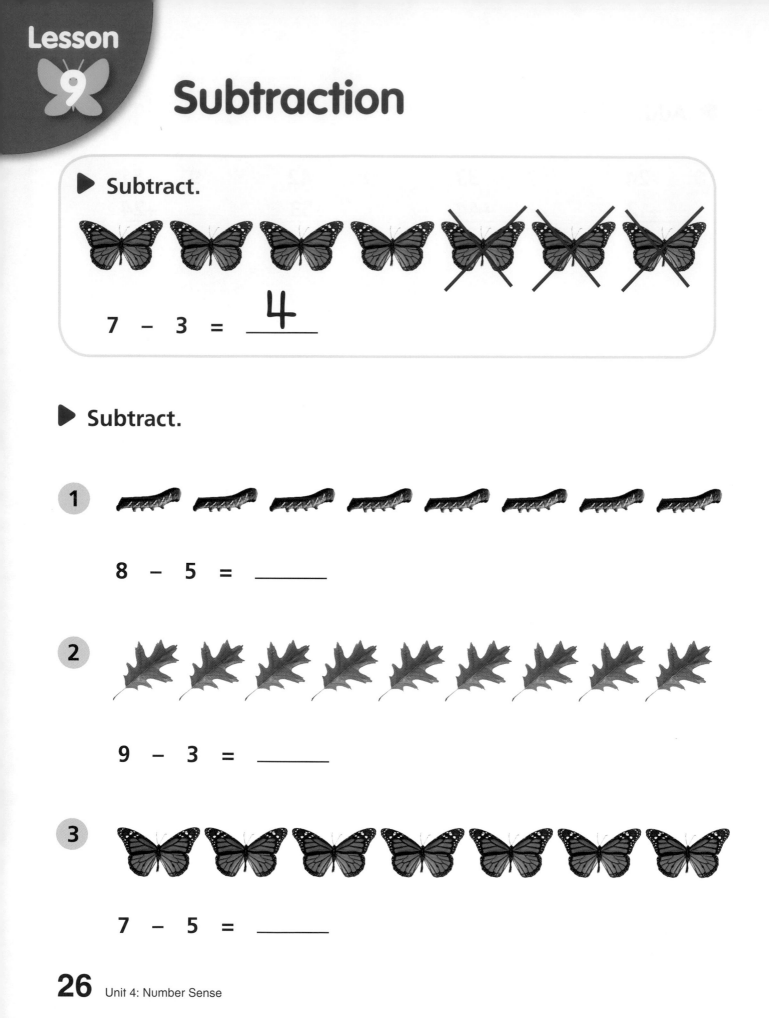

7 – 3 = __4__

▶ **Subtract.**

1

8 – 5 = _____

2

9 – 3 = _____

3

7 – 5 = _____

4

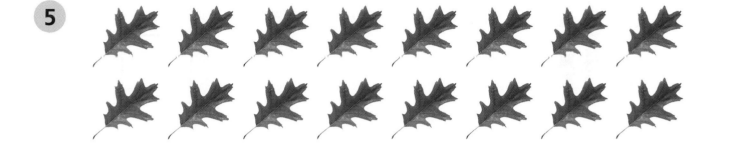

12 – 4 = _____

5

16 – 7 = _____

6

14 – 10 = _____

7

15 – 12 = _____

Subtract Tens and Ones

▶ Subtract. 37 − 13 = ?

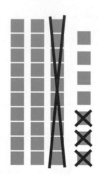

First, subtract the ones.
7 − 3 = 4.
Write 4 in the ones place.

Tens	Ones
3	7
− 1	3
	4

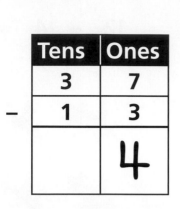

Then subtract the tens.
3 − 1 = 2
Write 2 in the tens place.

37 − 13 = _24_

Tens	Ones
3	7
− 1	3
2	4

▶ Subtract.

1 25 − 4 = ?

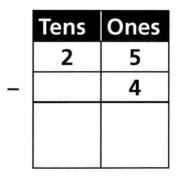

Tens	Ones
2	5
−	4

2 18 − 13 = ?

Tens	Ones
1	8
− 1	3

3 32 − 11 = ?

Tens	Ones
3	2
− 1	1

4   59 − 48 = ?

Tens	Ones
5	9
− 4	8

5   65 − 32 = ?

Tens	Ones
6	5
− 3	2

6   76 − 31 = ?

Tens	Ones
7	6
− 3	1

7   54 − 13 = ?

Tens	Ones
5	4
− 1	3

8   49 − 26 = ?

Tens	Ones
4	9
− 2	6

9   87 − 52 = ?

Tens	Ones
8	7
− 5	2

10   64
     − 42

11   73
     − 21

12   88
     − 46

13   95
     − 61

Subtraction with Regrouping

► **Regroup to subtract. 32 − 17 = ?**

Regroup one ten as 10 ones.
10 + 2 = 12.

Tens	Ones
2	12
~~3~~	~~2~~
− 1	7

Now, subtract the ones.
12 − 7 = 5.
Write 5 in the ones place.

Tens	Ones
2	12
~~3~~	~~2~~
− 1	7
	5

Then subtract the tens.
2 − 1 = 1.
Write 1 in the tens place.

Tens	Ones
2	12
~~3~~	~~2~~
− 1	7
1	5

32 − 17 = _____

► **Regroup to subtract.**

1 35 − 19 = ?

Tens	Ones
3	5
− 1	9

2 42 − 25 = ?

Tens	Ones
4	2
− 2	5

3 56 − 19 = ?

Tens	Ones
5	6
− 1	9

4 53 − 24 = ?

Tens	Ones
5	3
− 2	4

5 67 − 28 = ?

Tens	Ones
6	7
− 2	8

6 71 − 34 = ?

Tens	Ones
7	1
− 3	4

7
```
  43
− 27
```

8
```
  56
− 19
```

9
```
  74
− 65
```

10
```
  81
− 39
```

Checkup

▶ **Subtract.**

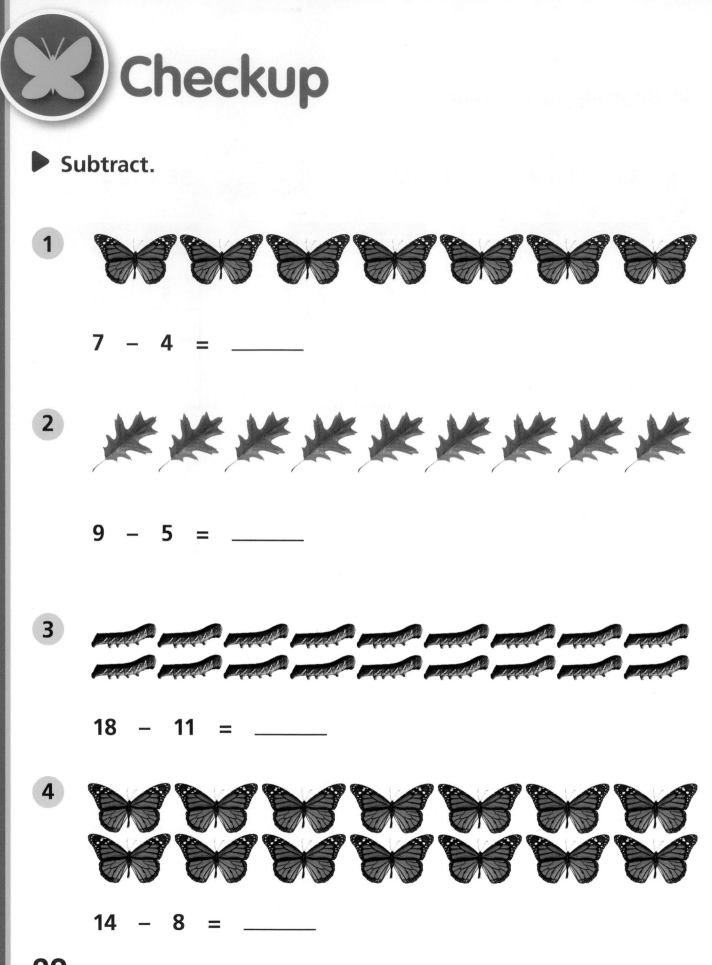

1

7 – 4 = _____

2

9 – 5 = _____

3

18 – 11 = _____

4

14 – 8 = _____

▶ **Subtract.**

5 27 – 6 = ?

Tens	Ones
2	7
–	6

6 49 – 14 = ?

Tens	Ones
4	9
– 1	4

7 58 – 24 = ?

Tens	Ones
5	8
– 2	4

8   53
  – 12

9   66
  – 43

10   89
  – 42

11   95
  – 23

▶ **Regroup to subtract.**

12 43 – 18 = ?

Tens	Ones
4	3
– 1	8

13 64 – 47 = ?

Tens	Ones
6	4
– 4	7

14   81
  – 35

Add Equal Groups

▶ Count the number of groups. Count the number in each group. Find how many in all.

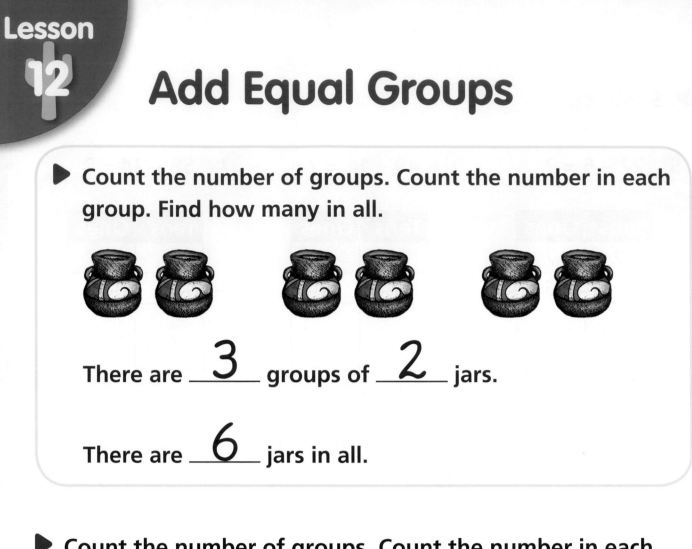

There are ___3___ groups of ___2___ jars.

There are ___6___ jars in all.

▶ Count the number of groups. Count the number in each group. Find how many in all.

1

There are _____ groups of _____ ears of corn.

There are _____ ears of corn in all.

2

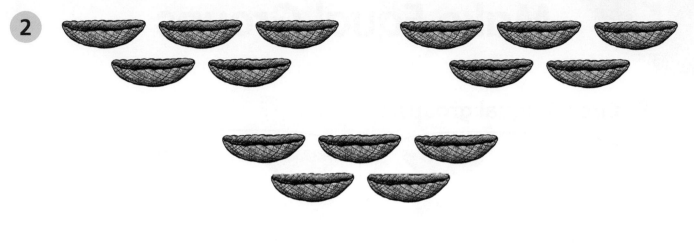

There are _____ groups of _____ baskets.

There are _____ baskets in all.

3

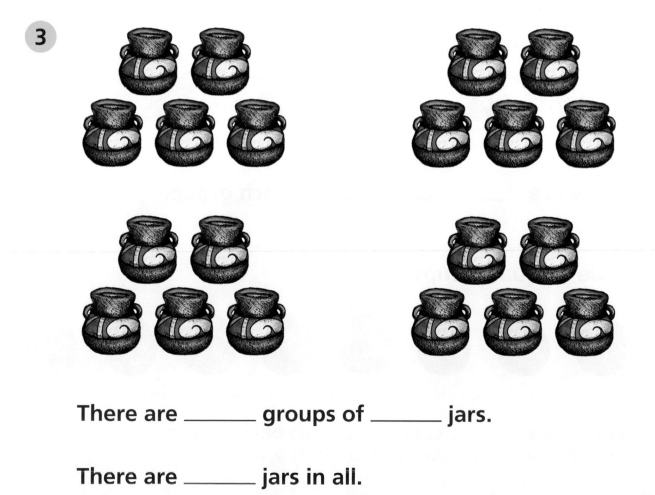

There are _____ groups of _____ jars.

There are _____ jars in all.

Make Equal Groups

▶ Circle 2 equal groups.

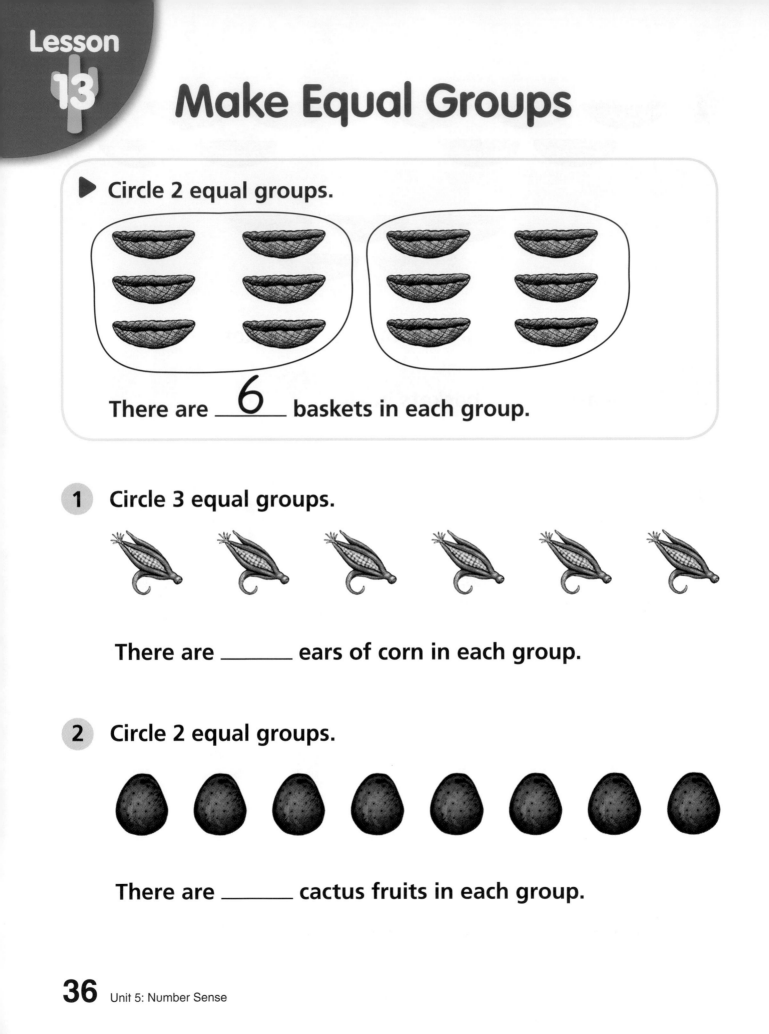

There are ___6___ baskets in each group.

1 Circle 3 equal groups.

There are _____ ears of corn in each group.

2 Circle 2 equal groups.

There are _____ cactus fruits in each group.

3 Circle 5 equal groups.

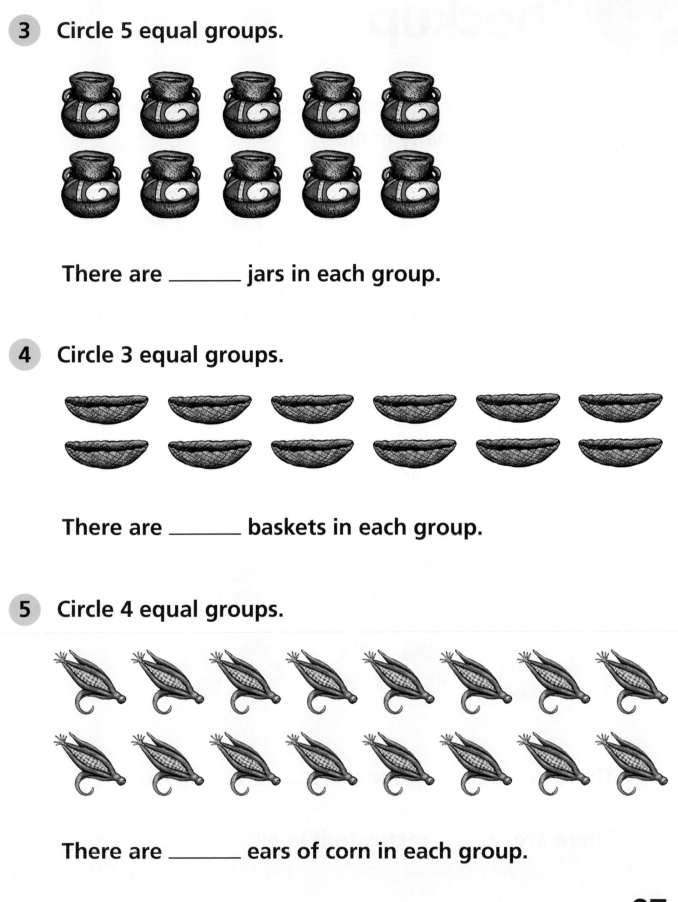

There are _____ jars in each group.

4 Circle 3 equal groups.

There are _____ baskets in each group.

5 Circle 4 equal groups.

There are _____ ears of corn in each group.

 # Checkup

▶ **Count the number of groups. Count the number in each group. Find how many in all.**

1

There are _____ groups of _____ jars.

There are _____ jars in all.

2

There are _____ groups of _____ cactus fruit.

There are _____ cactus fruit in all.

38 Unit 5: Number Sense

3 Circle 3 equal groups.

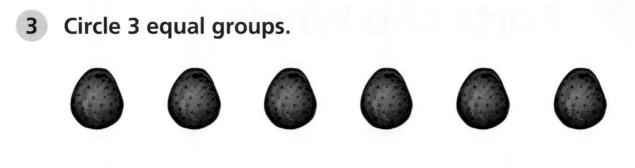

There are _____ cactus fruits in each group.

4 Circle 5 equal groups.

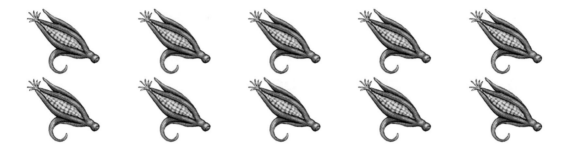

There are _____ ears of corn in each group.

5 Circle 4 equal groups.

There are _____ jars in each group.

Parts of a Whole

▶ **Write the fraction for the shaded part.**

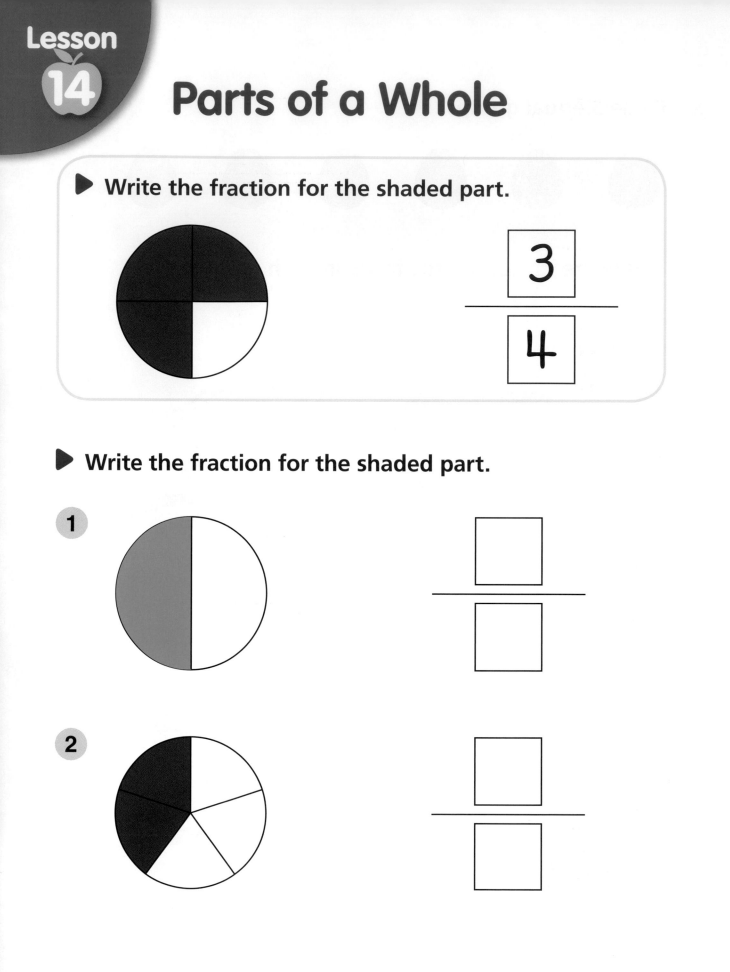

$$\frac{3}{4}$$

▶ **Write the fraction for the shaded part.**

1

2

3

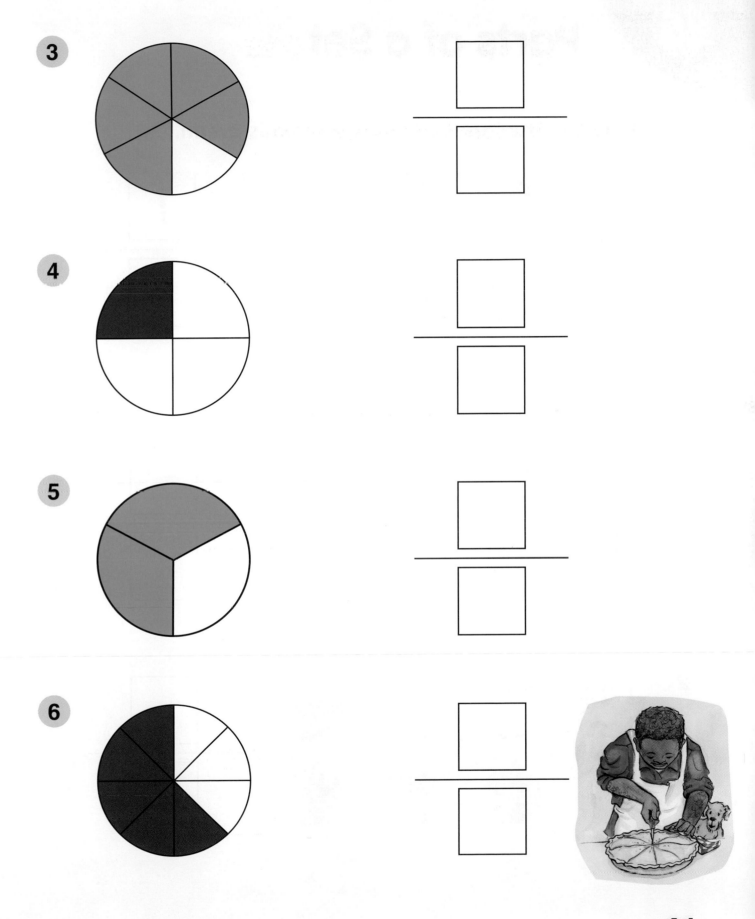

4

5

6

Parts of a Set

▶ **Write the fraction of the group that is green.**

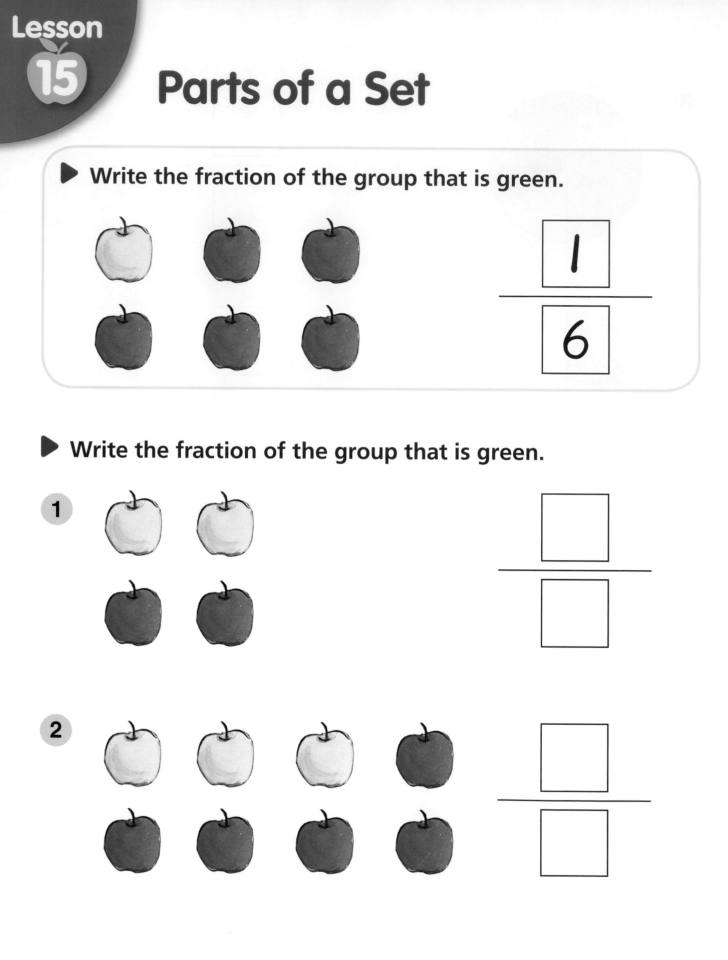

$$\frac{1}{6}$$

▶ **Write the fraction of the group that is green.**

1

2

▶ **Write the fraction of the group that is orange.**

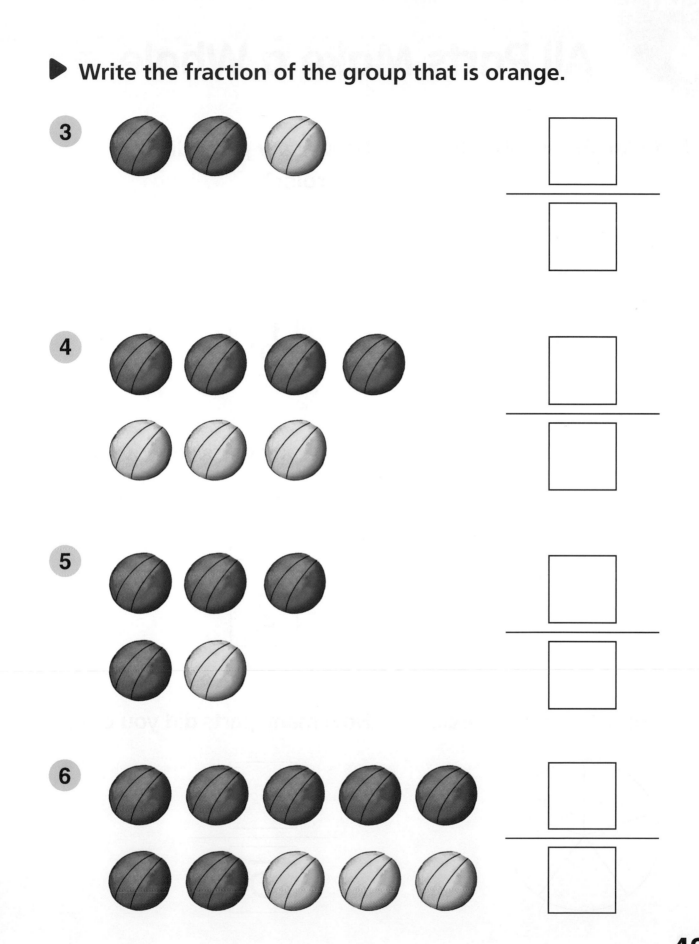

3

4

5

6

All Parts Make a Whole

▶ Color the whole circle.   How many parts did you color?

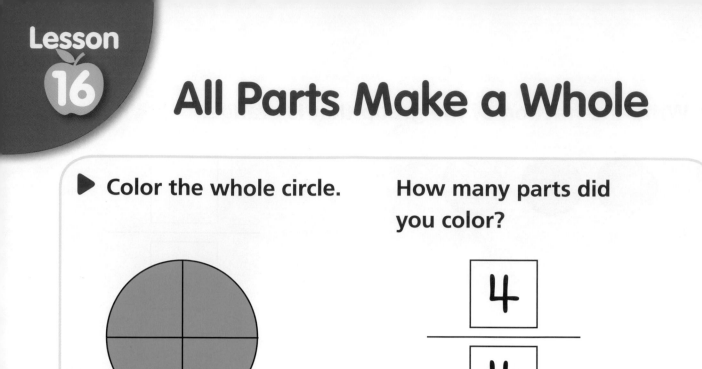

$$\frac{4}{4}$$

1 Color the whole circle.   How many parts did you color?

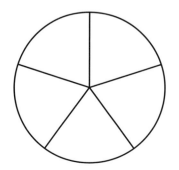

$$\frac{}{2}$$

2 Color the whole circle.   How many parts did you color?

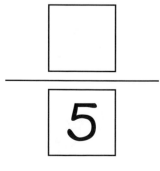

$$\frac{}{5}$$

3 Color the whole circle.   How many parts did you color?

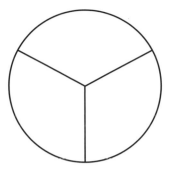

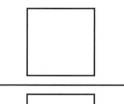
$$\frac{}{3}$$

4 Color the whole circle.   How many parts did you color?

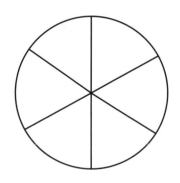

$$\frac{}{6}$$

5 Color the whole circle.   How many parts did you color?

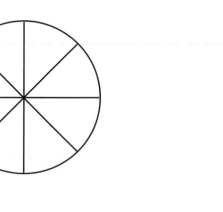

$$\frac{}{8}$$

▶ **Write the fraction for the shaded part.**

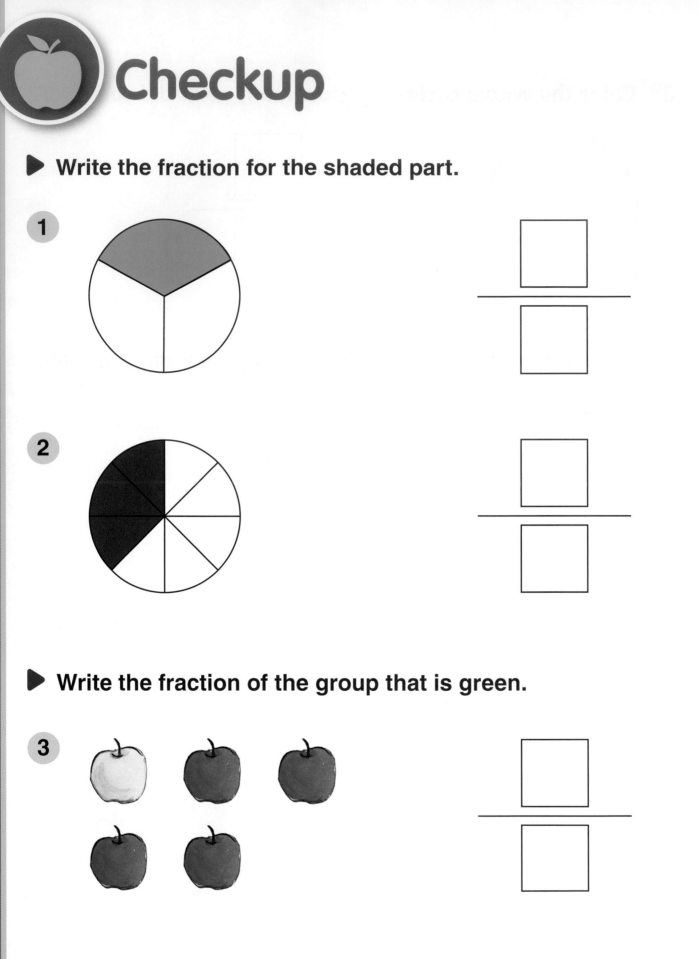

1

2

▶ **Write the fraction of the group that is green.**

3

4 Write the fraction of the group that is orange.

$$\frac{}{}$$

5 Color the whole circle. How many parts did you color?

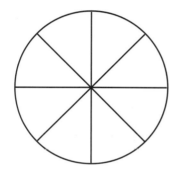

$$\frac{}{8}$$

6 Color the whole circle. How many parts did you color?

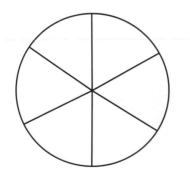

$$\frac{}{6}$$

Money

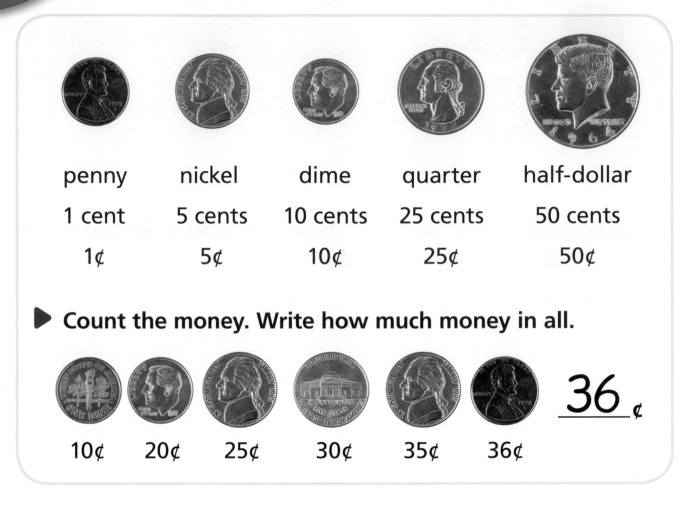

penny	nickel	dime	quarter	half-dollar
1 cent	5 cents	10 cents	25 cents	50 cents
1¢	5¢	10¢	25¢	50¢

▶ **Count the money. Write how much money in all.**

10¢   20¢   25¢   30¢   35¢   36¢   __36__ ¢

▶ **Count the money. Write how much money in all.**

1 _____ ¢

2 _____ ¢

© Options Publishing

3 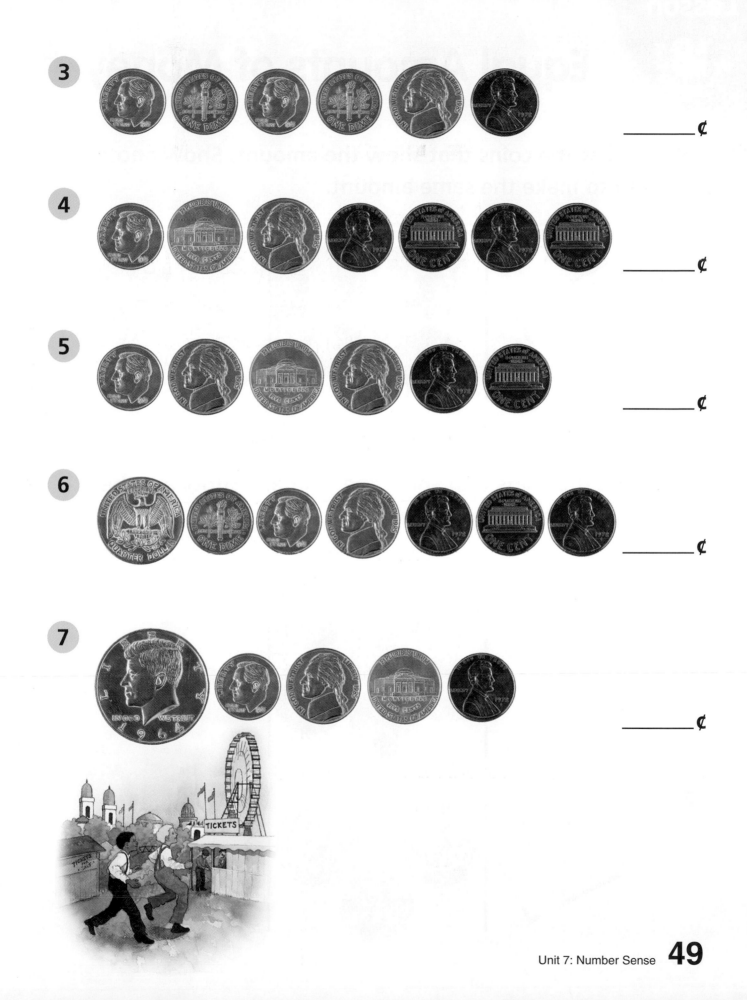 _____ ¢

4 _____ ¢

5 _____ ¢

6 _____ ¢

7 _____ ¢

Lesson 18

Equal Amounts of Money

▶ Look at the coins that show the amount. Show another way to make the same amount.

▶ Look at the coins that show the amount. Show another way to make the same amount.

1

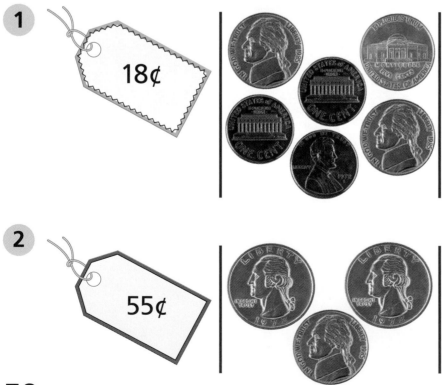

18¢

2

55¢

3

45¢

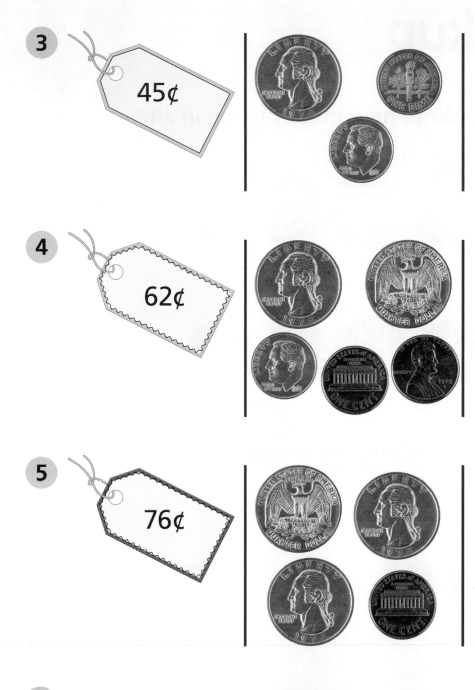

4

62¢

5

76¢

6

81¢

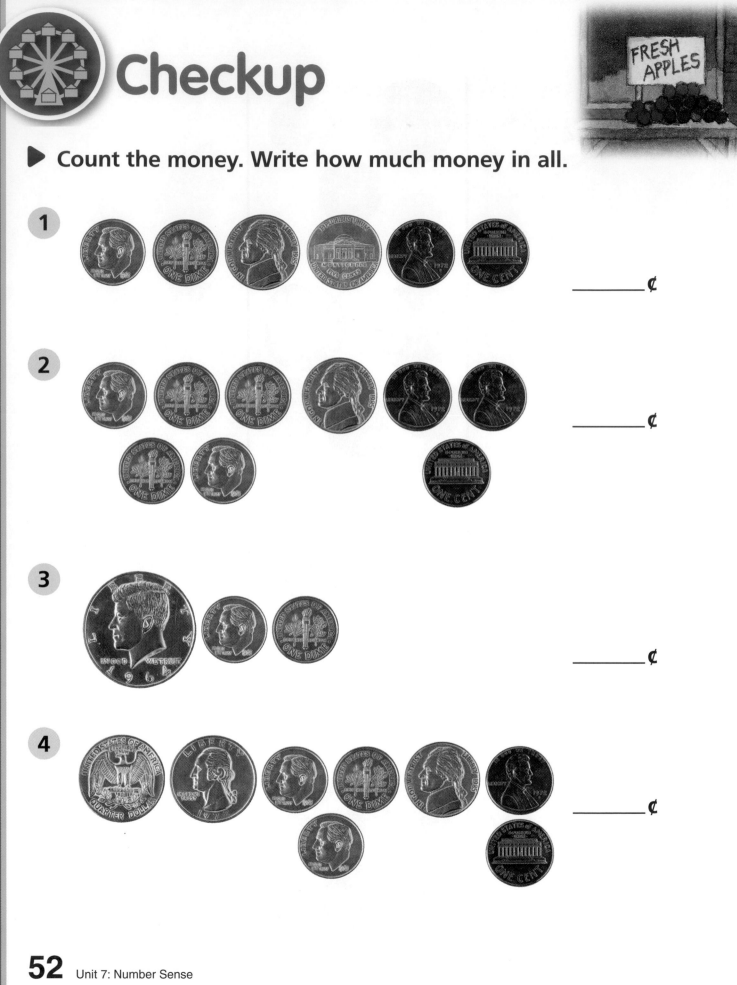

Checkup

FRESH APPLES

▶ **Count the money. Write how much money in all.**

1

_____ ¢

2

_____ ¢

3

_____ ¢

4

_____ ¢

▶ **Look at the coins that show the amount. Show another way to make the same amount.**

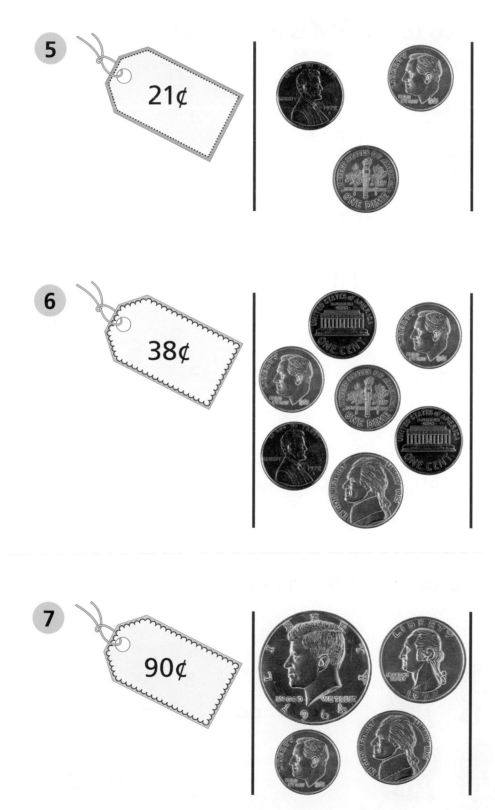

5 21¢

6 38¢

7 90¢

Even and Odd Numbers

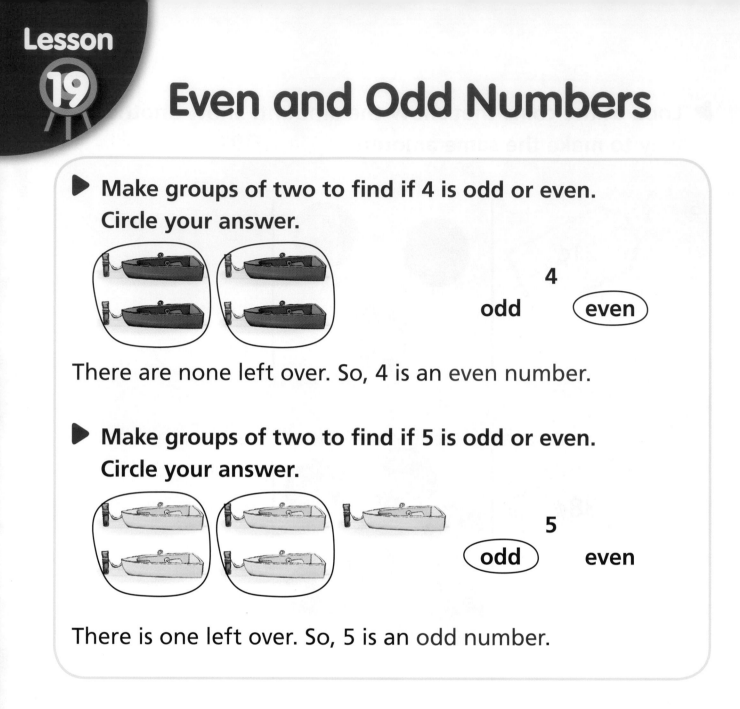

▶ **Make groups of two to find if 4 is odd or even. Circle your answer.**

4

odd   (even)

There are none left over. So, 4 is an even number.

▶ **Make groups of two to find if 5 is odd or even. Circle your answer.**

5

(odd)   even

There is one left over. So, 5 is an odd number.

▶ **Make groups of two to find if the number is odd or even. Circle your answer.**

1

8

odd   even

2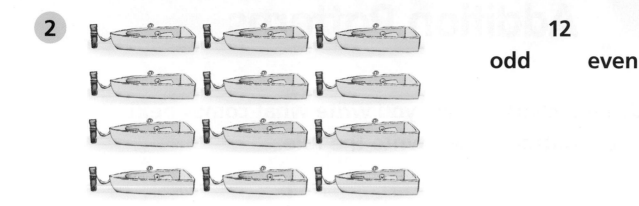

12

odd   even

3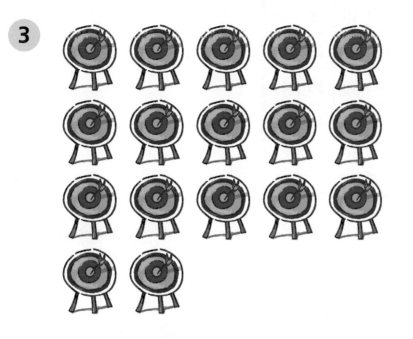

17

odd   even

4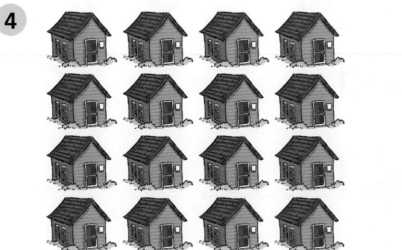

16

odd   even

Addition Patterns

▶ **Use the chart to help you write what comes next in the pattern. Then write the rule.**

1	2	3	4	5	6	7	8	9	10
11	12	13	14	15	16	17	18	19	20
21	22	23	24	25	26	27	28	29	30
31	32	33	34	35	36	37	38	39	40
41	42	43	44	45	46	47	48	49	50
51	52	53	54	55	56	57	58	59	60
61	62	63	64	65	66	67	68	69	70
71	72	73	74	75	76	77	78	79	80
81	82	83	84	85	86	87	88	89	90
91	92	93	94	95	96	97	98	99	100

12　14　16　18　_20_, _22_

The rule is ___add 2___.

2　4　6　8　10　12　_14_, _16_, _18_

The rule is _____.

▶ **Use the chart to help you write what comes next in the pattern. Then write the rule.**

1 | 5 | 10 | 15 | 20 | 25 | _____, _____, _____

The rule is _____.

2 | 10 | 20 | 30 | _____, _____, _____, _____

The rule is _____.

3 | 55 | 60 | 65 | _____, _____, _____, _____

The rule is _____.

4 | 78 | 80 | 82 | _____, _____, _____, _____

The rule is _____.

5 | 24 | _____, _____, | 30 | 32 | 34 |

The rule is _____.

Subtraction Patterns

▶ Use the chart to help you write what comes next in the pattern. Then write the rule.

1	2	3	4	5	6	7	8	9	10
11	12	13	14	15	16	17	18	19	20
21	22	23	24	25	26	27	28	29	30
31	32	33	34	35	36	37	38	39	40
41	42	43	44	45	46	47	48	49	50
51	52	53	54	55	56	57	58	59	60
61	62	63	64	65	66	67	68	69	70
71	72	73	74	75	76	77	78	79	80
81	82	83	84	85	86	87	88	89	90
91	92	93	94	95	96	97	98	99	100

40 38 36 34   __32__, __30__

The rule is __subtract 2__.

83 82 81 80   __79__, __78__

The rule is _____.

▶ **Use the chart to help you write what comes next in the pattern. Then write the rule.**

1 70 65 60 55 50 _____, _____, _____

The rule is _____.

2 70 60 50 40 _____, _____, _____

The rule is _____.

3 100 98 96 94 _____, _____, _____

The rule is _____.

4 44 43 _____, _____, _____, 39 38

The rule is _____.

5 80 70 _____, _____, 40 30

The rule is _____.

Paired Numbers

You can use a table to find a pattern.
You know that a cat has 2 eyes.

So, 2 cats have ___4___ eyes, and 3 cats have ___6___ eyes.

▶ **Continue the pattern to find the answer.**
How many eyes do five cats have?

Number of cats	1	2	3	4	5
Number of eyes	2	4	6	8	10

▶ **Continue the pattern to find the answer.**

1 **How many wheels do 5 tricycles have?**

Number of tricycles	1	2	3	4	5
Number of wheels	3	6	___	___	___

Five tricycles have _____ wheels.

2 How many wheels do 6 wagons have?

Number of	1	2	3	4	5	6
Number of wheels	4	8	12	_____	_____	_____

Six wagons have _____ wheels.

3 How many pennies equal 5 nickels?

Number of nickels	1	2	3	4	5
Number of pennies	5	10	_____	_____	_____

_____ pennies equal five nickels.

4 How many pennies equal 5 dimes?

Number of dimes	1	2	3	4	5
Number of pennies	10	20	_____	_____	_____

_____ pennies equal five dimes.

Checkup

▶ **Make groups of two to find if the number is odd or even. Circle your answer.**

1

10

odd   even

▶ **Use the chart to help you write what comes next in the pattern. Then write the rule.**

1	2	3	4	5	6	7	8	9	10
11	12	13	14	15	16	17	18	19	20
21	22	23	24	25	26	27	28	29	30
31	32	33	34	35	36	37	38	39	40
41	42	43	44	45	46	47	48	49	50
51	52	53	54	55	56	57	58	59	60
61	62	63	64	65	66	67	68	69	70
71	72	73	74	75	76	77	78	79	80
81	82	83	84	85	86	87	88	89	90
91	92	93	94	95	96	97	98	99	100

2 35 40 45 50 55 _____, _____, _____

The rule is _____.

3 | 57 | 56 | 55 | 54 | _____, _____, _____

The rule is _____.

▶ Continue the pattern to find the answer.

4 How many legs do 5 dogs have?

Number of dogs	1	2	3	4	5
Number of legs	4	8	___	___	___

Five dogs have _____ legs.

5 How many legs do 6 birds have?

Number of birds	1	2	3	4	5	6
Number of legs	2	4	___	___	___	___

Six birds have _____ legs.

Addition Number Sentences

▶ Find the missing number. Use the number line to help you.

$6 + \triangle = 8$

1   2

0   1   2   3   4   5   6   7   8

The missing number is __2__.

▶ Find the missing number. Use the number line to help you.

1 $8 + \blacksquare = 12$

0   1   2   3   4   5   6   7   8   9   10   11   12

The missing number is _____.

2 $7 + \bullet = 10$

0   1   2   3   4   5   6   7   8   9   10

The missing number is _____.

3 9 + ▲ = 13

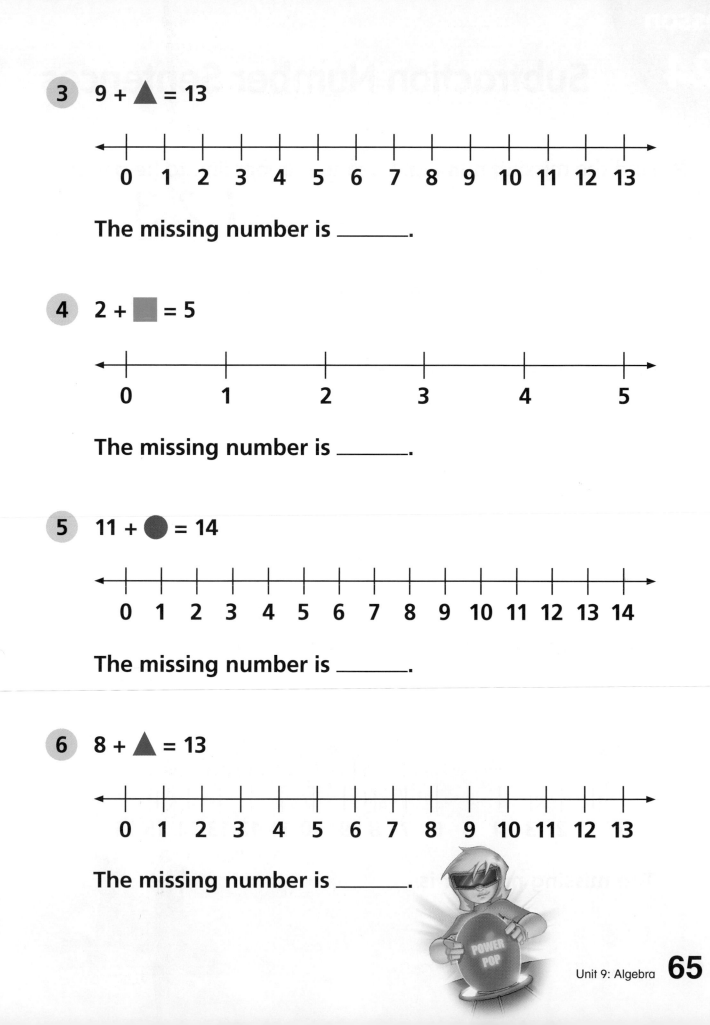

The missing number is _____.

4 2 + ■ = 5

The missing number is _____.

5 11 + ● = 14

The missing number is _____.

6 8 + ▲ = 13

The missing number is _____.

Subtraction Number Sentences

▶ **Find the missing number. Use the number line to help you.**

12 − ■ = 9

1 2 3

0 1 2 3 4 5 6 7 8 9 10 11 12

The missing number is __3__.

▶ **Find the missing number. Use the number line to help you.**

1 11 − ● = 6

0 1 2 3 4 5 6 7 8 9 10 11

The missing number is _____.

2 16 − ▲ = 12

0 1 2 3 4 5 6 7 8 9 10 11 12 13 14 15 16

The missing number is _____.

3 $13 - \blacksquare = 8$

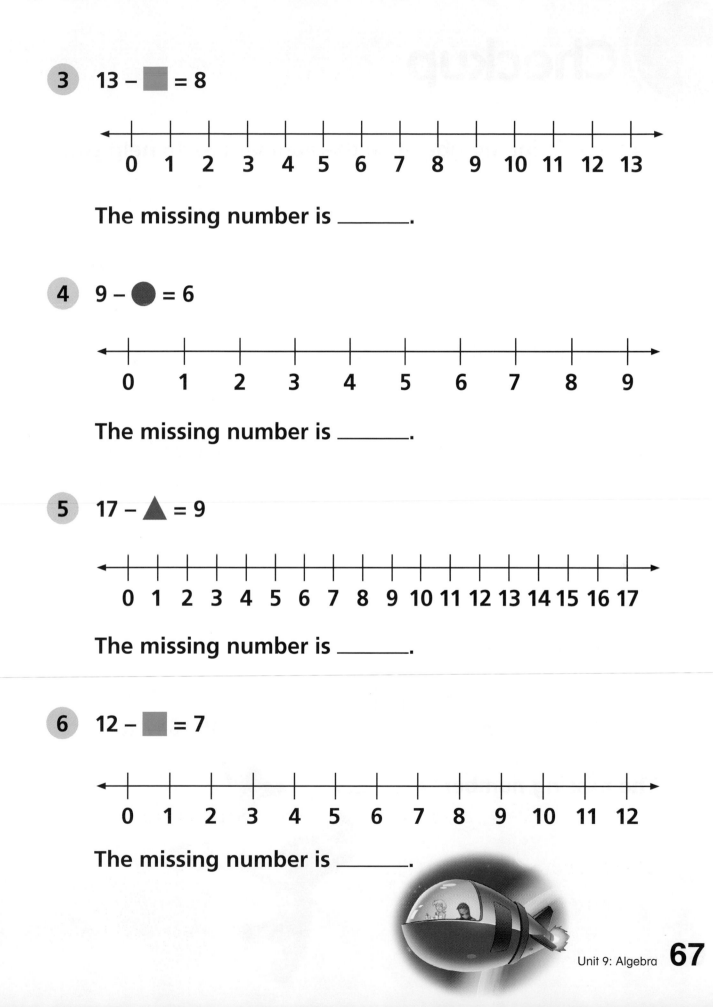

0 1 2 3 4 5 6 7 8 9 10 11 12 13

The missing number is _____.

4 $9 - \bullet = 6$

0 1 2 3 4 5 6 7 8 9

The missing number is _____.

5 $17 - \blacktriangle = 9$

0 1 2 3 4 5 6 7 8 9 10 11 12 13 14 15 16 17

The missing number is _____.

6 $12 - \blacksquare = 7$

0 1 2 3 4 5 6 7 8 9 10 11 12

The missing number is _____.

Checkup

▶ **Find the missing number. Use the number line to help you.**

1 4 + ● = 7

```
←——+——+——+——+——+——+——+——→
   0   1   2   3   4   5   6   7
```

The missing number is _____.

2 9 + ▲ = 11

```
←——+——+——+——+——+——+——+——+——+——+——+——→
   0   1   2   3   4   5   6   7   8   9  10  11
```

The missing number is _____.

3 8 + ■ = 18

```
←+—+—+—+—+—+—+—+—+—+—+—+—+—+—+—+—+—+——→
 0 1 2 3 4 5 6 7 8 9 10 11 12 13 14 15 16 17 18
```

The missing number is _____.

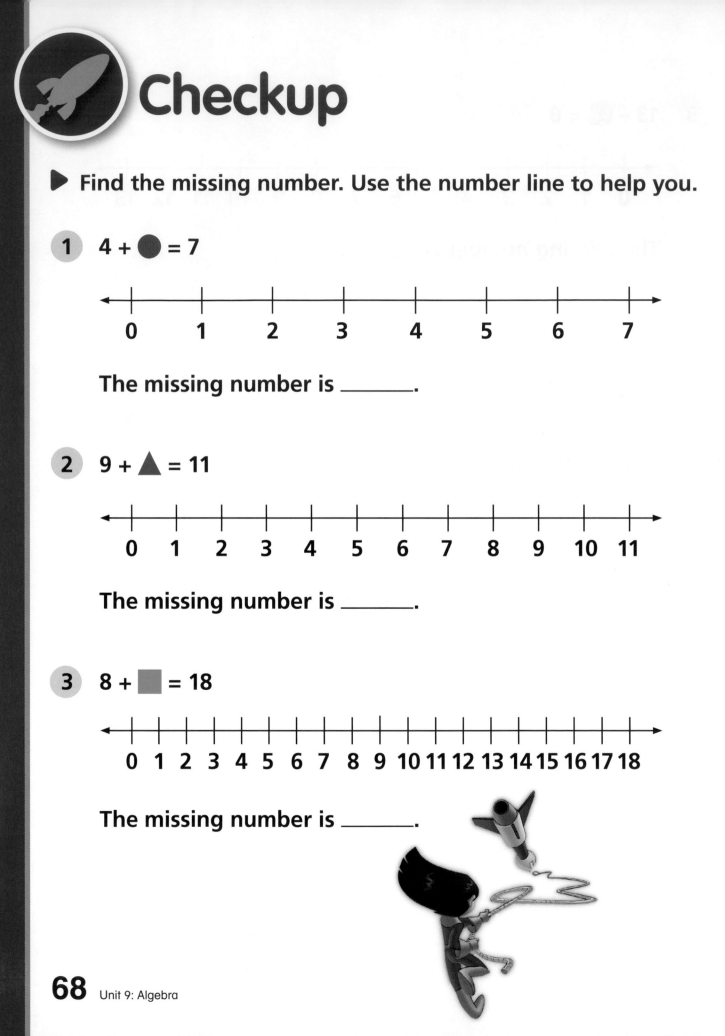

4 $12 - \bullet = 9$

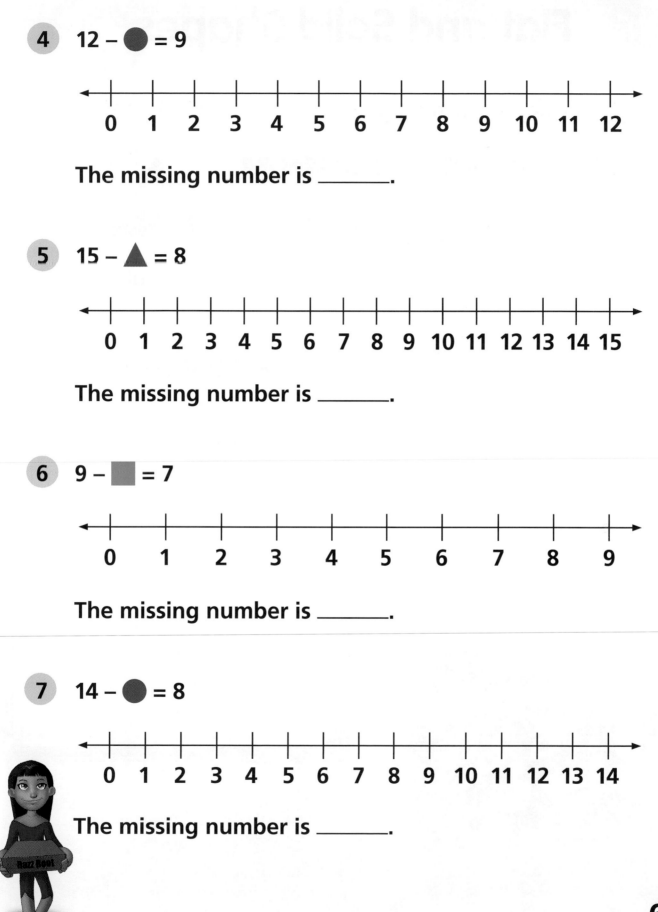

The missing number is _____.

5 $15 - \blacktriangle = 8$

The missing number is _____.

6 $9 - \blacksquare = 7$

The missing number is _____.

7 $14 - \bullet = 8$

The missing number is _____.

Flat and Solid Shapes

▶ **Look at these flat shapes.**

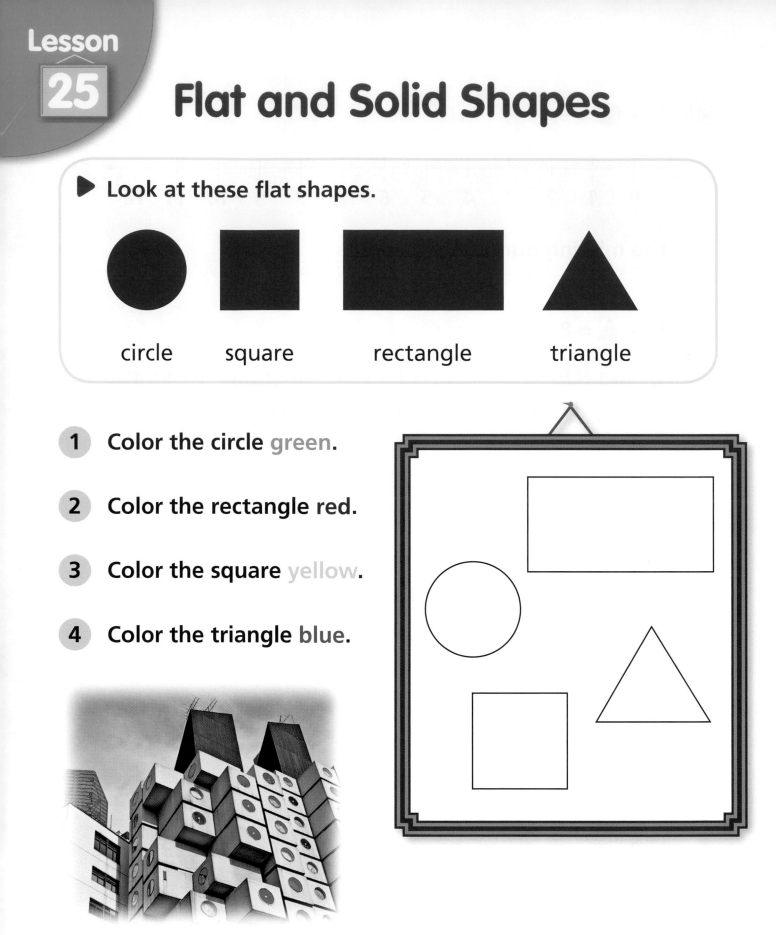

circle   square   rectangle   triangle

1. **Color the circle green.**

2. **Color the rectangle red.**

3. **Color the square yellow.**

4. **Color the triangle blue.**

► **Look at these solid shapes.**

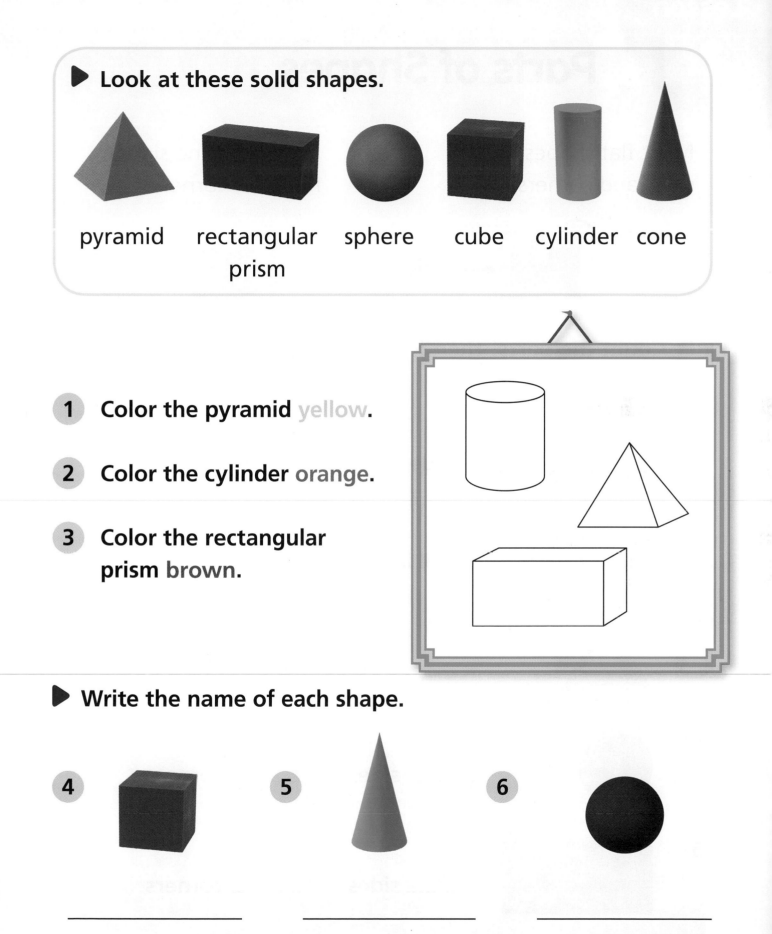

pyramid   rectangular prism   sphere   cube   cylinder   cone

1   **Color the pyramid** yellow.

2   **Color the cylinder** orange.

3   **Color the rectangular prism** brown.

► **Write the name of each shape.**

4   **5**   **6**

_____   _____   _____

Parts of Shapes

Most flat shapes have sides and corners.

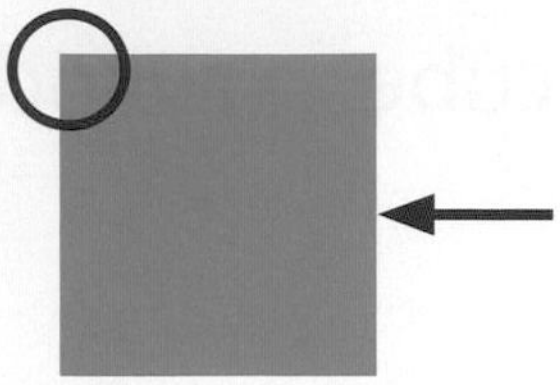

A circle has no sides. It has no corners.

corner ◯ ← side

circle

▶ **Count the parts. Then fill in the blanks.**

__4__ sides   __4__ corners

▶ **Count the parts. Then fill in the blanks.**

1 _____ sides   _____ corners

2 _____ sides   _____ corners

3 _____ sides   _____ corners

A face is a flat side on a shape. An edge is where two faces meet. A vertex is a corner where 3 or more faces meet. The plural of vertex is vertices.

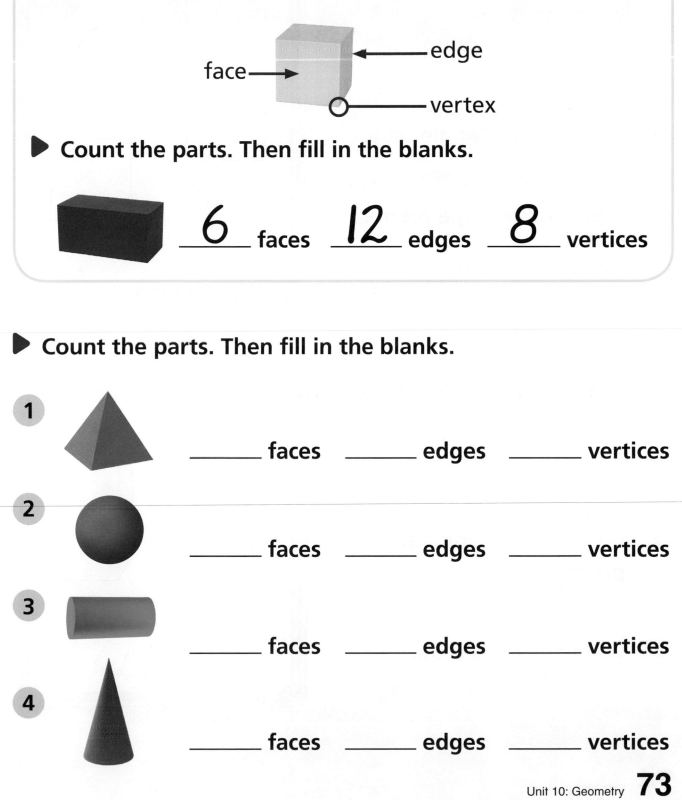

face——▶
edge
vertex

▶ **Count the parts. Then fill in the blanks.**

__6__ faces   __12__ edges   __8__ vertices

▶ **Count the parts. Then fill in the blanks.**

1
_____ faces   _____ edges   _____ vertices

2
_____ faces   _____ edges   _____ vertices

3
_____ faces   _____ edges   _____ vertices

4
_____ faces   _____ edges   _____ vertices

Checkup

1 Color the circle red.

2 Color the square purple.

3 Color the rectangle blue.

4 Color the triangle orange.

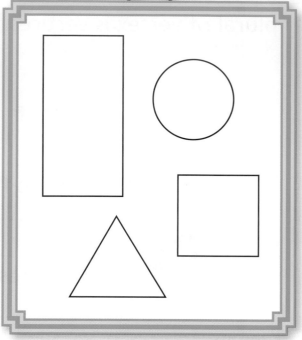

5 Color the cone green.

6 Color the pyramid orange.

7 Color the rectangular prism yellow.

8 Color the cube blue.

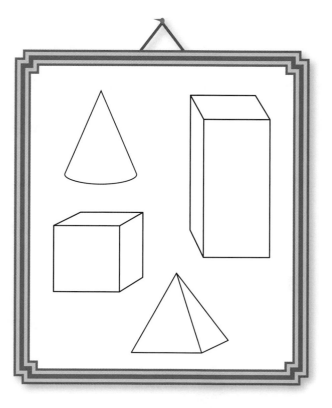

74 Unit 10: Geometry

▶ **Count the parts. Then fill in the blanks.**

9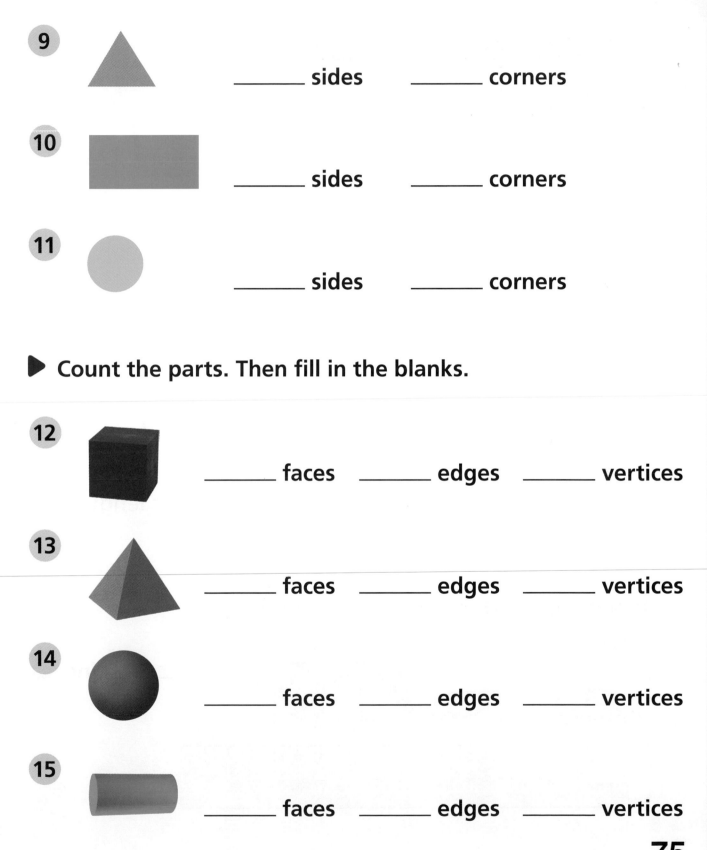
　　　　　　_____ sides　　　_____ corners

10
　　　　　　_____ sides　　　_____ corners

11
　　　　　　_____ sides　　　_____ corners

▶ **Count the parts. Then fill in the blanks.**

12
　　　　　　_____ faces　　_____ edges　　_____ vertices

13
　　　　　　_____ faces　　_____ edges　　_____ vertices

14
　　　　　　_____ faces　　_____ edges　　_____ vertices

15
　　　　　　_____ faces　　_____ edges　　_____ vertices

Inches and Feet

▶ **How many inches long is this frog?**

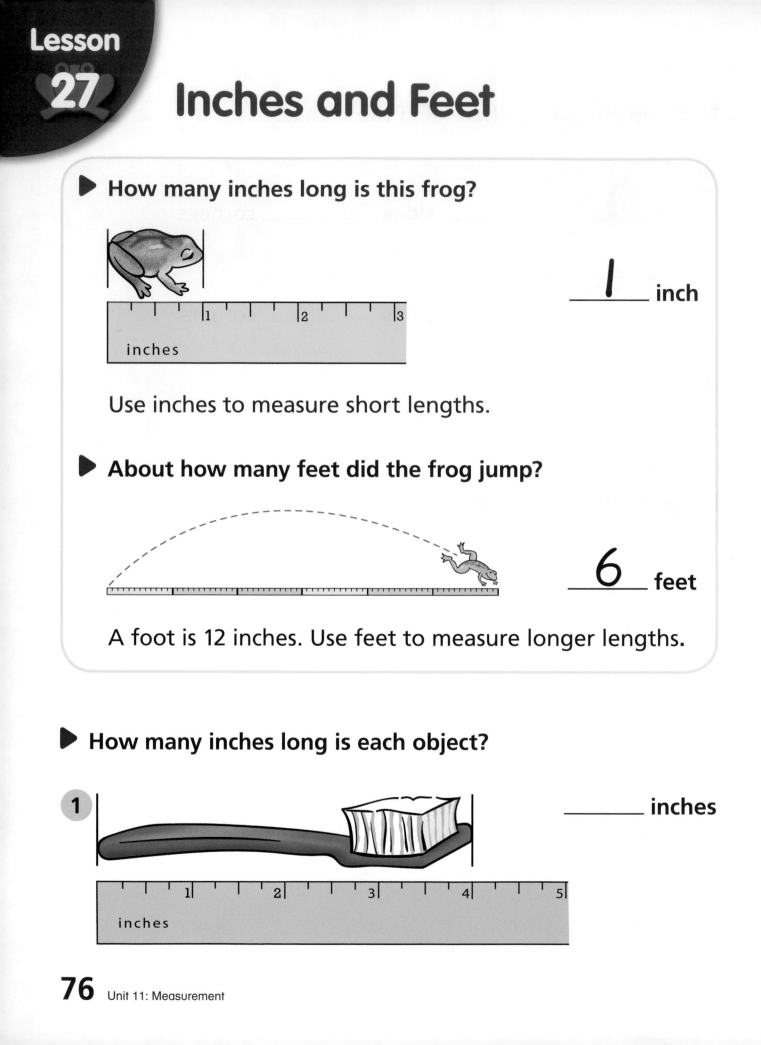

_____1_____ inch

Use inches to measure short lengths.

▶ **About how many feet did the frog jump?**

_____6_____ feet

A foot is 12 inches. Use feet to measure longer lengths.

▶ **How many inches long is each object?**

1

_____ inches

2 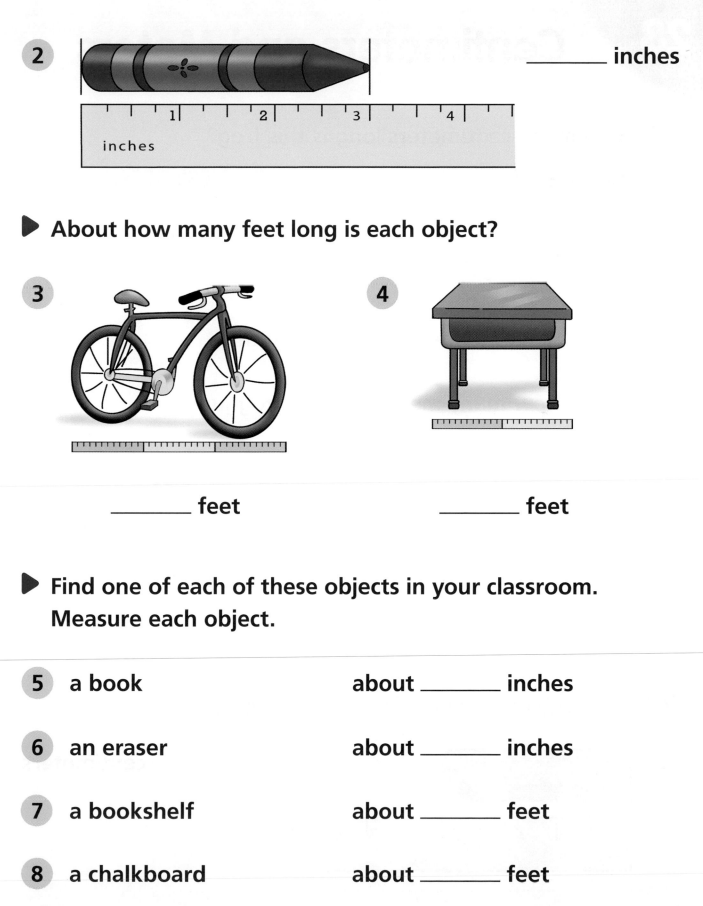 _____ inches

▶ **About how many feet long is each object?**

3

_____ feet

4

_____ feet

▶ **Find one of each of these objects in your classroom. Measure each object.**

5 a book   about _____ inches

6 an eraser   about _____ inches

7 a bookshelf   about _____ feet

8 a chalkboard   about _____ feet

Centimeters and Meters

▶ **How many centimeters long is this frog?**

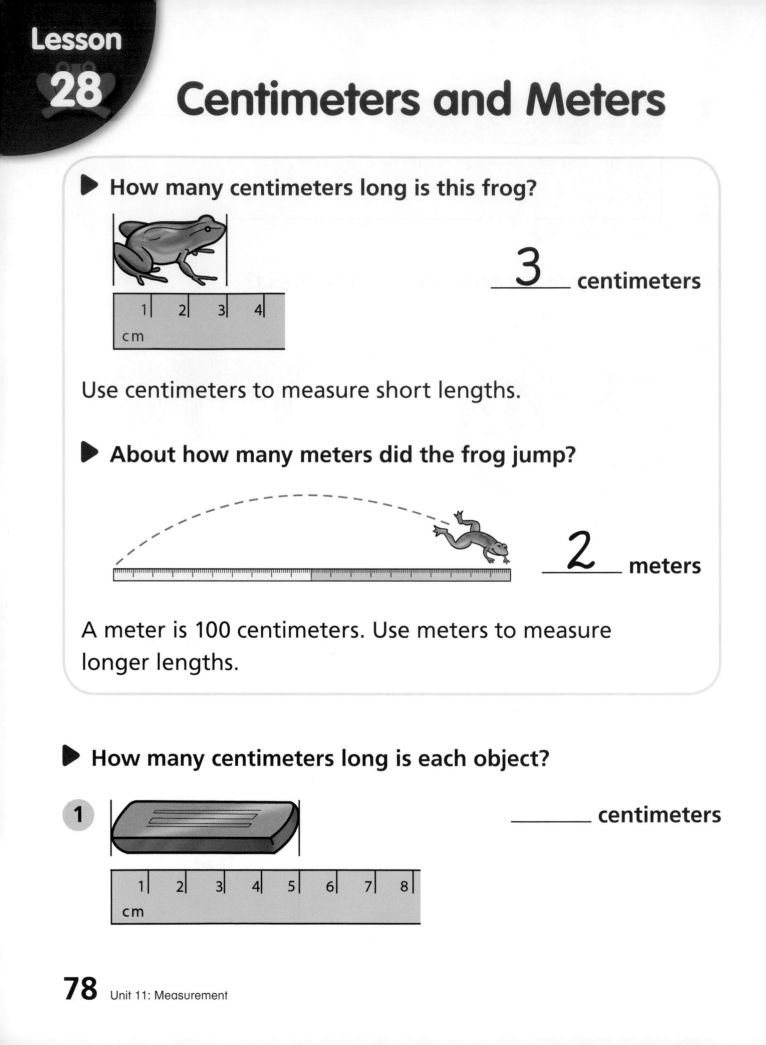

_____ **3** centimeters

Use centimeters to measure short lengths.

▶ **About how many meters did the frog jump?**

_____ **2** meters

A meter is 100 centimeters. Use meters to measure longer lengths.

▶ **How many centimeters long is each object?**

1 _____ centimeters

2 _____ centimeters

```
1   2   3   4   5   6   7   8   9
cm
```

▶ **About how many meters long is each object?**

3

_____ meters

4

_____ meters

▶ **Find one of each of these objects in your classroom. Measure each object.**

5 a piece of notebook paper   about _____ centimeters

6 a pair of scissors   about _____ centimeters

7 a table   about _____ meters

8 a student desk   about _____ meters

Ounces and Pounds

You can find the weight of
an object in ounces or pounds.
There are 16 ounces in 1 pound.

▶ **Circle how much each object weighs.**

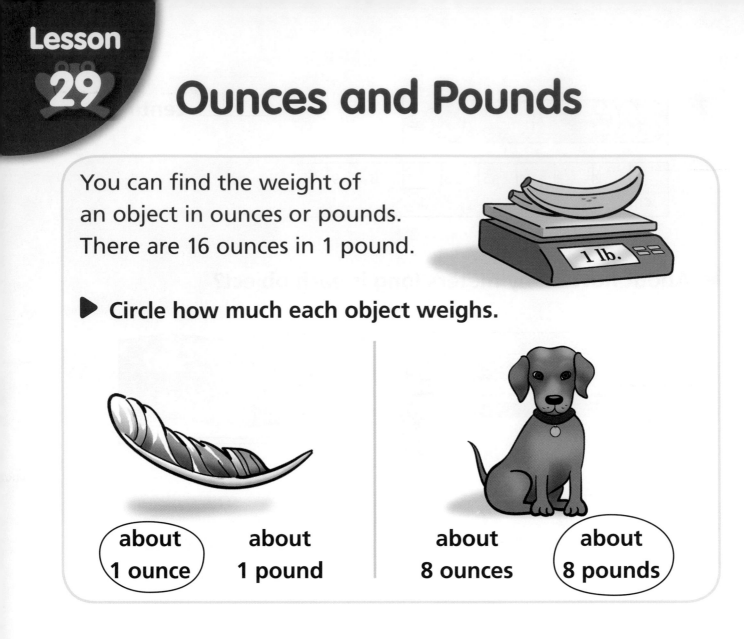

(about 1 ounce)   about 1 pound

about 8 ounces   (about 8 pounds)

▶ **Circle how much each object weighs.**

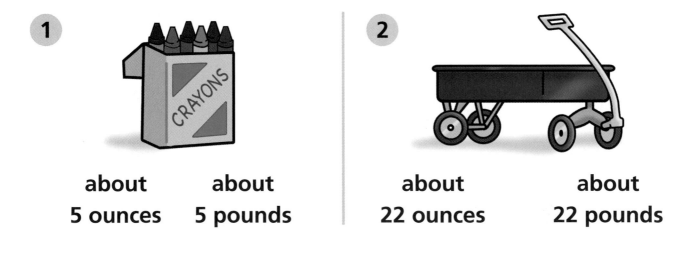

1

about 5 ounces   about 5 pounds

2

about 22 ounces   about 22 pounds

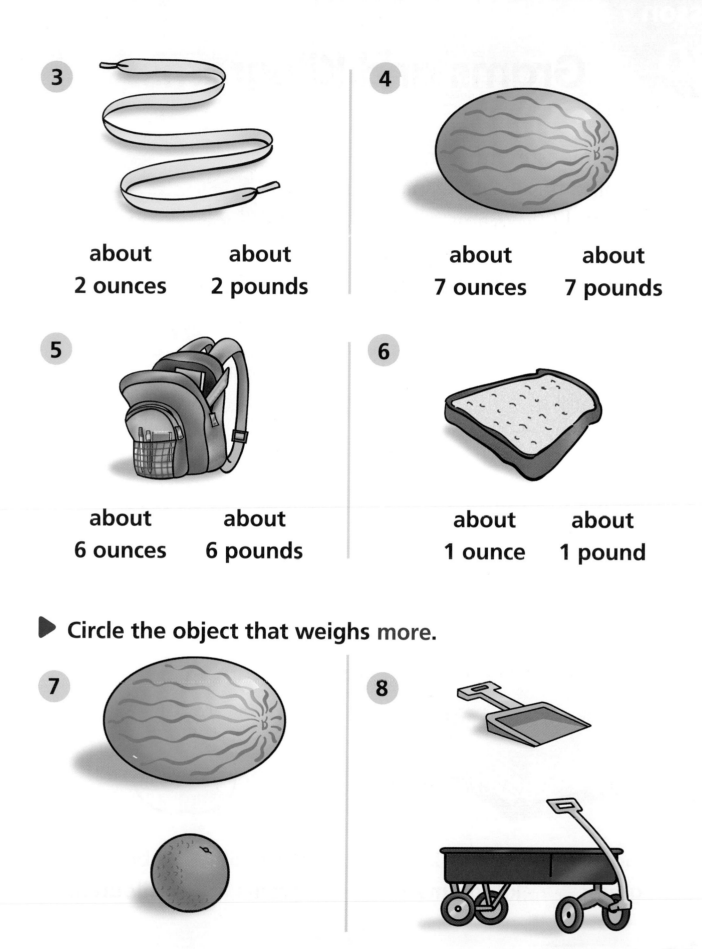

3

about
2 ounces

about
2 pounds

4

about
7 ounces

about
7 pounds

5

about
6 ounces

about
6 pounds

6

about
1 ounce

about
1 pound

▶ **Circle the object that weighs more.**

7

8

Grams and Kilograms

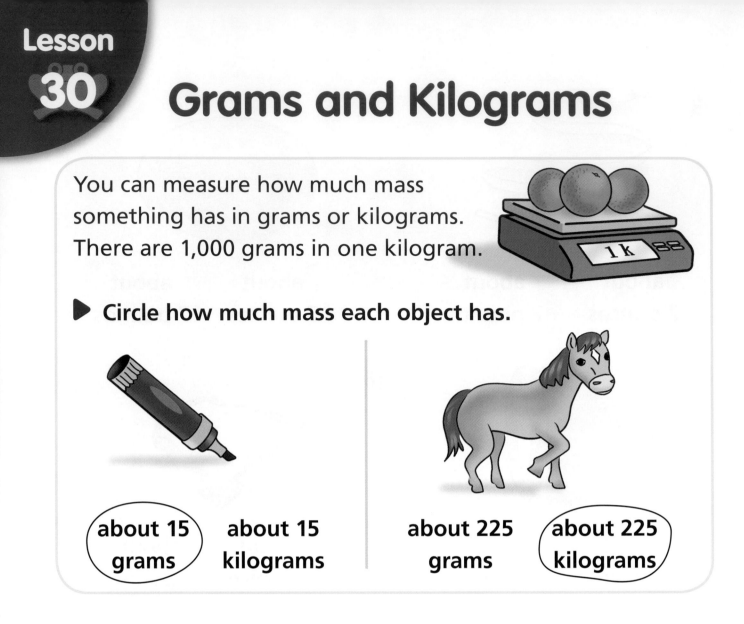

You can measure how much mass something has in grams or kilograms. There are 1,000 grams in one kilogram.

▶ Circle how much mass each object has.

about 15 grams

about 15 kilograms

about 225 grams

about 225 kilograms

▶ Circle how much mass each object has.

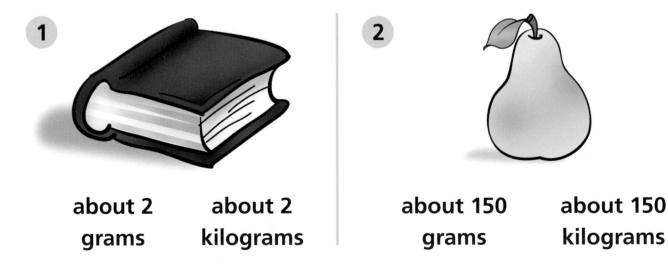

1

about 2 grams

about 2 kilograms

2

about 150 grams

about 150 kilograms

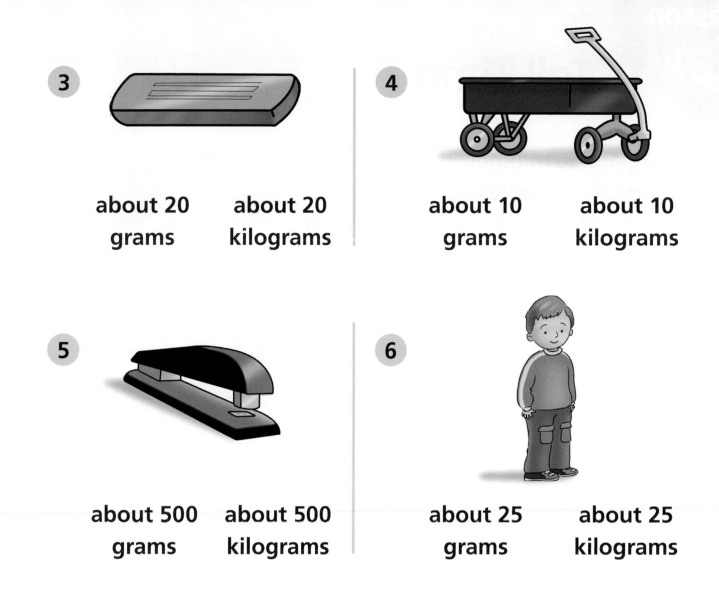

3 about 20 grams   about 20 kilograms

4 about 10 grams   about 10 kilograms

5 about 500 grams   about 500 kilograms

6 about 25 grams   about 25 kilograms

▶ Circle the object that has more mass.

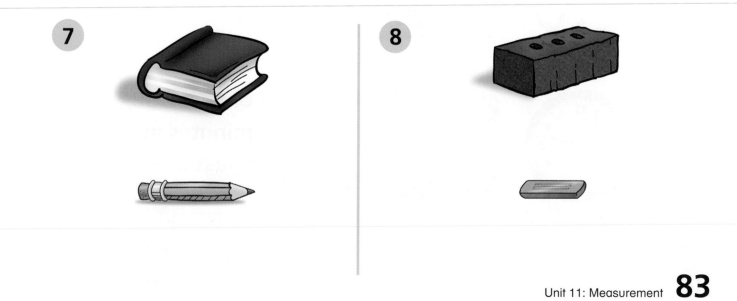

7

8

Tell Time

There are 60 minutes in one hour. There are 5 minutes between each number on a clock.

▶ **Read the clock. Write the time.**

1  **2:05**

Write or say: 2:05, or 5 minutes after 2.

2  **2:15**

Write or say: 2:15, or quarter past 2.

3

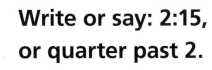

Write or say: 2:30, thirty minutes after 2, or half past 2.

▶ Read the clock. Write the time.

1
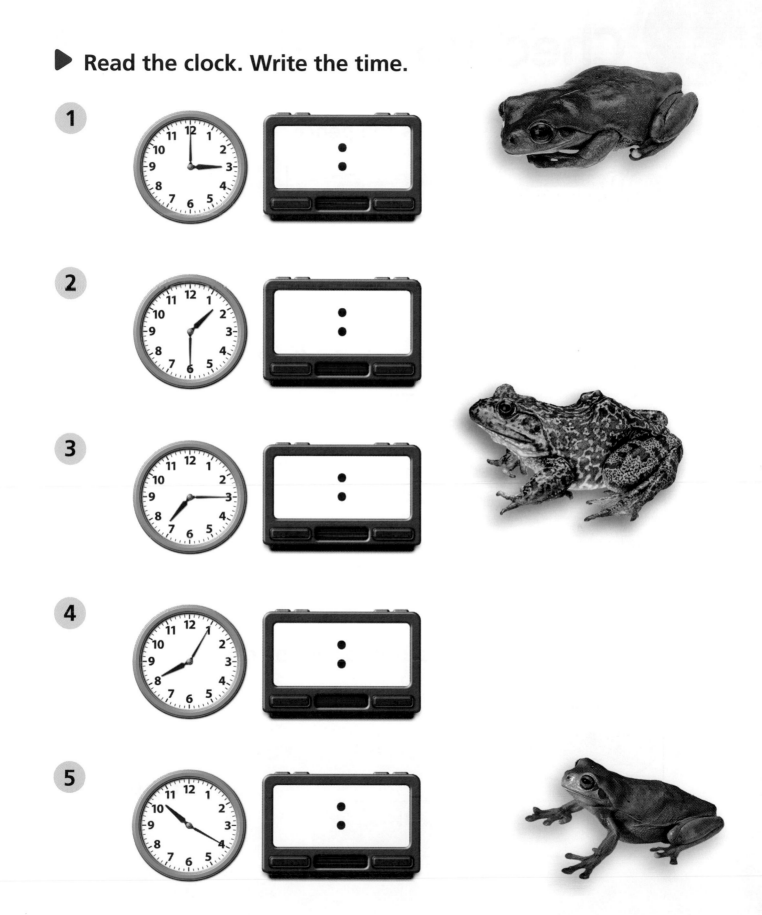

2

3

4

5

© Options Publishing

Checkup

1 How many inches long is the pencil?

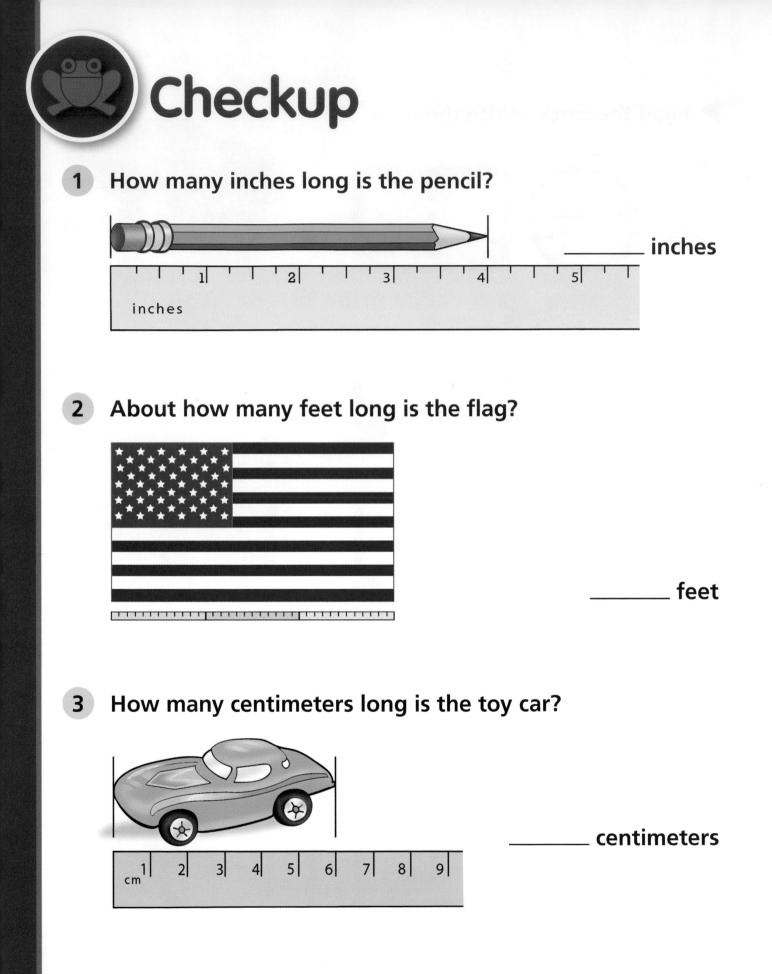

_____ inches

2 About how many feet long is the flag?

_____ feet

3 How many centimeters long is the toy car?

_____ centimeters

4 About how many meters long is the board?

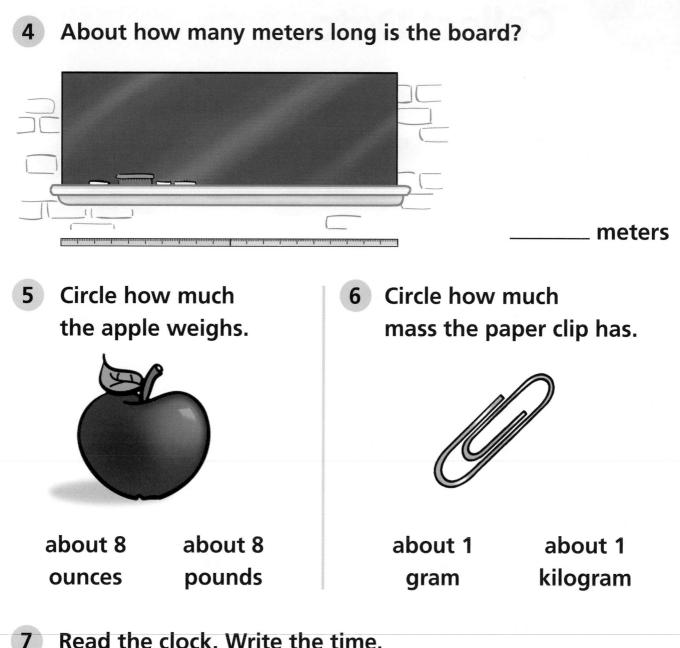

_____ meters

5 Circle how much the apple weighs.

about 8 ounces   about 8 pounds

6 Circle how much mass the paper clip has.

about 1 gram   about 1 kilogram

7 Read the clock. Write the time.

Collect Data

John asked some friends to vote for their favorite dinosaur. Each picture stands for one vote.

▶ **Put the votes in this tally chart.**

Favorite Dinosaur

Dinosaur	Votes
(T-Rex)	卌 I
(Brachiosaurus)	I
(Triceratops)	

▶ **How many friends voted for (T-Rex)?** _6_

▶ **How many friends voted for (Brachiosaurus)?** ____

A teacher asked students to vote for their favorite dinosaur. Each picture stands for one vote.

1 Put the votes in this tally chart.

Favorite Dinosaur

Dinosaur	Votes
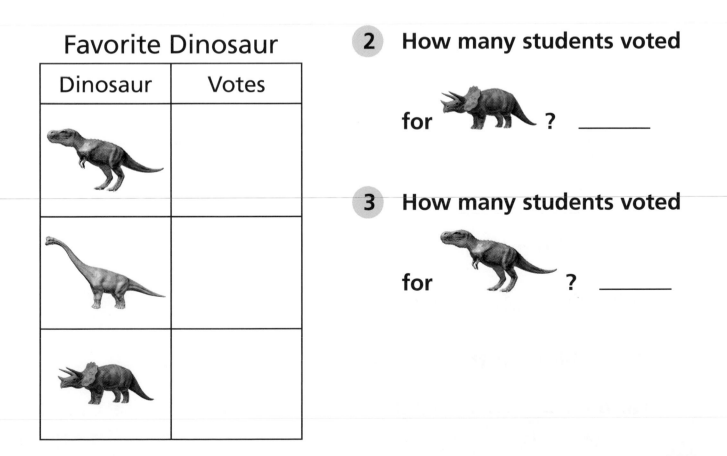	

2 How many students voted for ? _____

3 How many students voted for ? _____

Picture Graphs

Joni asked some friends to vote for their favorite bird.
She put their votes in this tally chart.

▶ **Use the tally chart to finish the picture graph.**

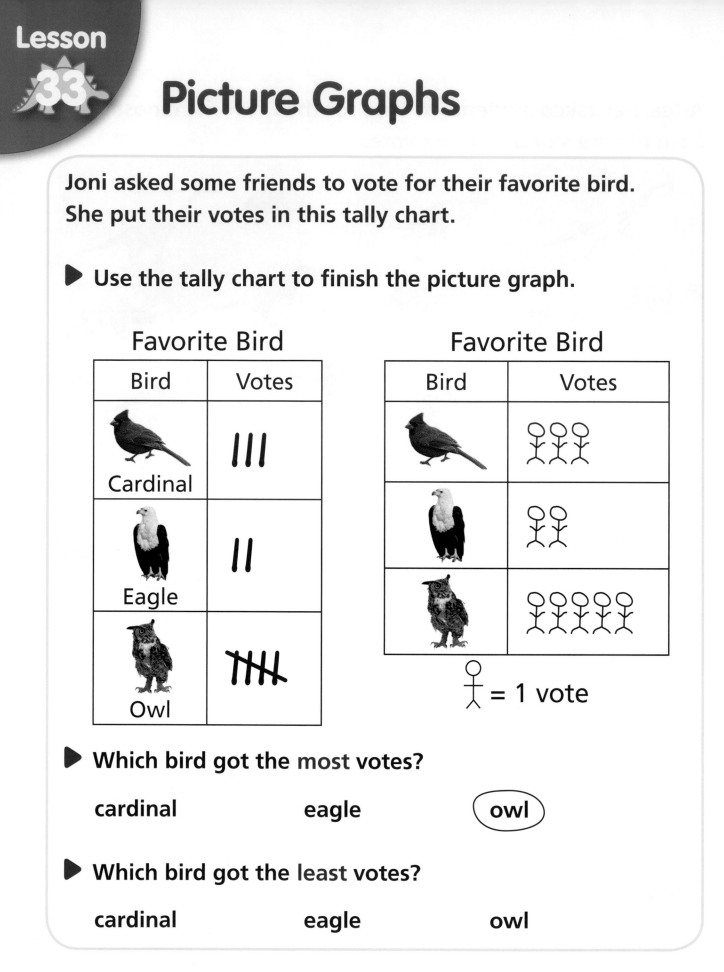

Favorite Bird

Bird	Votes
Cardinal	III
Eagle	II
Owl	卌

Favorite Bird

Bird	Votes

= 1 vote

▶ **Which bird got the most votes?**

cardinal   eagle   (owl)

▶ **Which bird got the least votes?**

cardinal   eagle   owl

Mark asked some friends to vote for their favorite fruit. He put their votes in this tally chart.

1 Use the tally chart to finish the picture graph.

Favorite Fruit

Fruit	Votes
Apple	ＨＨ l
Orange	ll
Banana	lll

Favorite Fruit

Fruit	Votes

⚘ = 1 vote

2 Which fruit got the most votes?

apple   orange   banana

3 Which fruit got the least votes?

apple   orange   banana

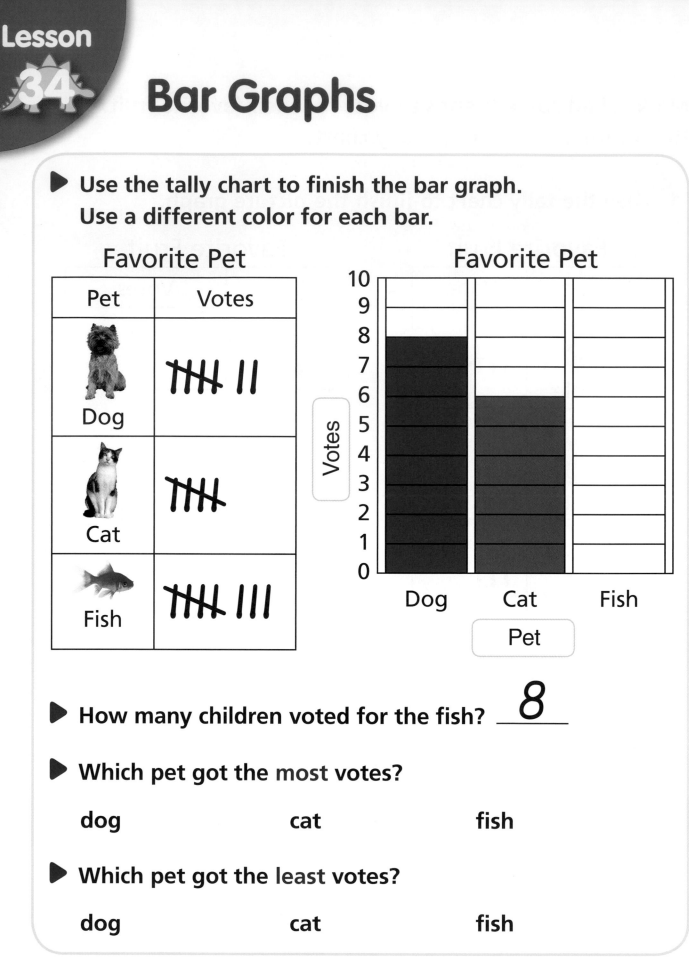

Bar Graphs

▶ Use the tally chart to finish the bar graph.
Use a different color for each bar.

Favorite Pet

Pet	Votes
Dog	卌 II
Cat	卌
Fish	卌 III

Favorite Pet

▶ How many children voted for the fish? __8__

▶ Which pet got the most votes?

dog   cat   fish

▶ Which pet got the least votes?

dog   cat   fish

1 Use the tally chart to finish the bar graph.
Use a different color for each bar.

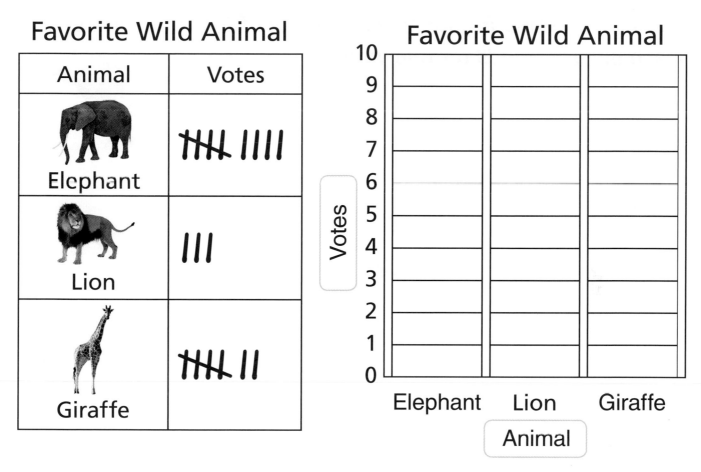

Favorite Wild Animal

Animal	Votes
Elephant	ⵜⵜⵜ IIII
Lion	III
Giraffe	ⵜⵜⵜ II

Favorite Wild Animal

Votes
10
9
8
7
6
5
4
3
2
1
0

Elephant   Lion   Giraffe

Animal

2 How many children voted for the giraffe? _____

3 Which animal got the **most votes**?

    elephant   **lion**   **giraffe**

4 Which animal got the **least votes**?

    elephant   **lion**   **giraffe**

Checkup

Lucy asked some friends to vote for their favorite dinosaur.
Each picture stands for one vote.

1 Put the votes in this tally chart.

Favorite Dinosaur

Dinosaur	Votes

2 How many friends voted

for ? _____

3 How many friends voted

for ? _____

Chrissy asked some friends to vote for their favorite reptile. She put their answers in this tally chart.

4 Use the tally chart to finish the picture graph.

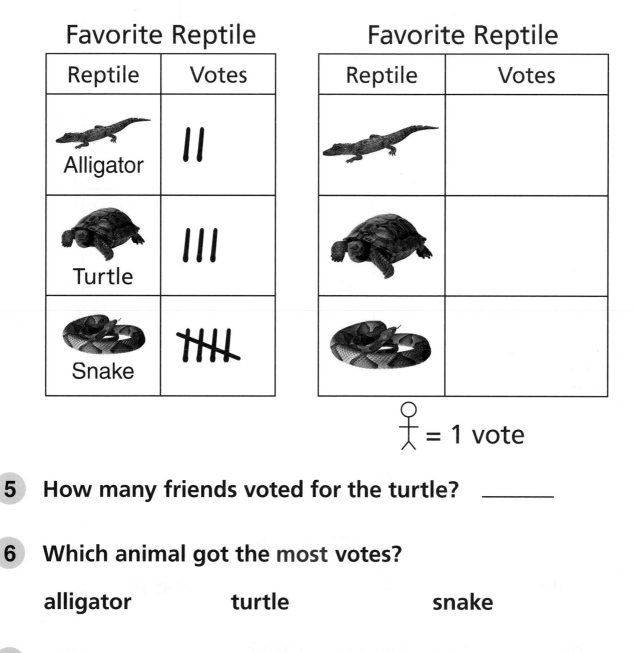

Favorite Reptile

Reptile	Votes
Alligator	||
Turtle	|||
Snake	ⅢⅠ

Favorite Reptile

Reptile	Votes

👤 = 1 vote

5 How many friends voted for the turtle? _____

6 Which animal got the most votes?

alligator   turtle   snake

7 Which animal got the least votes?

alligator   turtle   snake

8 Use the tally chart to finish the bar graph.
   Use a different color for each bar.

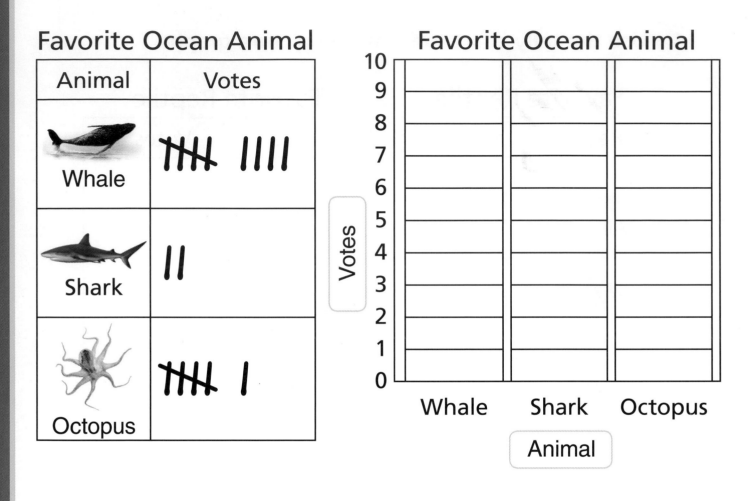

Favorite Ocean Animal

Animal	Votes
Whale	卌 IIII
Shark	II
Octopus	卌 I

Favorite Ocean Animal

Votes

Whale   Shark   Octopus

Animal

9 How many people voted for the shark? _____

10 Which animal got the **most** votes?

   whale   shark   octopus

11 Which animal got the **least** votes?

   whale   shark   octopus